NATURAL HISTORY OF THE BELL VIREO, VIREO BELLII AUDUBON

JON C. BARLOW

ZINC READ

Published by
This edition copyright © 2024 by Zinc Read
Design and setting By Zinc Read
Email- zincread@gmail.com
ISBN-978-93-6277-002-8

CONTENTS

INTRODUCTION

The Bell Vireo (*Vireo bellii* Aud.) is a summer resident in riparian and second growth situations in the central United States south of North Dakota. In the last two decades this bird has become fairly common in western, and to a lesser extent in central, Indiana and is apparently shifting its breeding range eastward in that state (Mumford, 1952; Nolan, 1960). In northeastern Kansas the species breeds commonly and occurs in most tracts of suitable habitat.

The literature contains several reports dealing exclusively with the Bell Vireo, notably those of Bennett (1917), Nice (1929), Du Bois (1940), Pitelka and Koestner (1942), Hensley (1950) and Mumford (1952). Bent (1950) has summarized the information available on the species through 1943. Nolan (1960) recently completed an extensive report based on a small, banded population at Bloomington, Monroe County, Indiana. He validated for this species many points of natural history previously based on estimates and approximations, especially concerning the post-fledging life of the young and the movement of the adults from one "home range"

to another in the course of a single season.

None of these reports, however, has emphasized the ritualized behavioral patterns associated with the maintenance of territory and with courtship. Among the North American Vireonidae, the behavior of the Red-eyed Vireo (*Vireo olivaceus*) is best documented (Sutton, 1949; Lawrence, 1953; Southern, 1958). With this

species authors have concentrated on the mechanics of the breeding season and their reports contain little discussion of the aggressive and epigamic behavior of the bird.

The amount of information on the ritualized behavior of the Bell Vireo and related species heretofore has been meager. I observed breeding behavior from its inception in early May through the summer of 1960. It is hoped the resulting information will serve as a basis of comparison in future studies of behavior of vireos; such ethological data are becoming increasingly important, especially as an aid in systematics.

ACKNOWLEDGMENTS

To professors Frank B. Cross, Henry S. Fitch, and Richard F. Johnston of the Department of Zoology of the University of Kansas I am grateful for comments and suggestions in various phases of the study and the preparation of the manuscript. Professor Johnston also made available data on the breeding of the Bell Vireo from the files of the Kansas Ornithological Society. I am indebted to my wife, Judith Barlow, for many hours of typing and copy reading.

Mrs. Lorna Cordonnier prepared the map, Thomas H. Swearingen drew the histograms, and Professor A. B. Leonard photographed and developed the histograms. Dr. Robert M. Mengel contributed the sketch of the Bell Vireo and George P. Young prepared the dummy Bell Vireo used in the field work. Thomas R. Barlow, Donald A. Distler, Abbot S. Gaunt, John L. Lenz, Gary L. Packard, A. Wayne Wiens, and John Wellman assisted in various phases of the field work.

METHODS OF STUDY

Daily observations were made from May 11 to June 26 in 1959 and from April 15 through July 15 in 1960. Six additional visits were made to the study area in September of 1959, and ten others in July and August, 1960. Periods of from one hour to eleven hours were spent in the field each day, and a total of about five hundred hours were logged in the field.

Each territory was visited for at least five minutes each day but more often for twenty minutes. The breeding activities of the pairs were rarely synchronous. Consequently several stages in the cycle of building were simultaneously available for study.

Nine young and one adult were banded in 1959. No Bell Vireos were banded in 1960. Individual pairs could be recognized because of their exclusive use of certain segments of the study area and by the individual variation in the song of the males. Sexes were distinguishable on the basis of differences in vocalizations and plumages.

Most nests were located by the observer searching, watching a pair engaged in building, or following a singing male until the increased tempo of his song indicated proximity to a nest. As the season progressed and the foliage grew more dense, it became increasingly difficult to locate completed nests. Blinds were unnecessary because of the density of vegetation. Observations were facilitated by a 7 x 50 binocular. Data were recorded on the spot in a field notebook. Eggs were numbered by means of Higgins Engrossing ink as they were laid.

Individual trees in which males sang most were marked over a three-week period. Then the distances between the most remote perches were paced. These distances aided in determining the size of the territories. The general configuration of the vegetation within each territory determined the location of one or more boundaries of the territory. Each territory was given a number, 1, 2, 3, etc., as it was discovered; consequently there is no numerical relationship between the designations of the territories established in 1959 and 1960. Nests within a territory were designated as 1-a, 1-b, 1-c, *etc.*

Although experimentation was not a primary source of data, it proved useful in certain instances. A stuffed Blue Jay elicited mobbing behavior from nesting pairs. A dummy Bell Vireo elicited both agonistic and epigamic behavior from nesting pairs, depending on the phase of the nesting cycle.

The temperature at the beginning of each day's work was taken by means of a Weston dial thermometer. A hand counter and a pocket watch having a second hand were used in determining such data as frequency of song and periods of attentiveness by the sexes.

Histological cross-sections, prepared by A. Wayne Wiens, of the ventral epidermis of both sexes were used to study brood patches.

STUDY AREA

The intensive field work was on a 39-acre tract (fig. 1) extending approximately 7/10 of a mile west from U. S. highway 59, which in 1959-1960 constituted the western city limit of Lawrence, Douglas County, Kansas. The eastern boundary of the study area is

approximately 1-1/2 miles southwest of the County Courthouse in Lawrence. The eastern ten acres is associated with the Laboratory of Aquatic Biology of the University of Kansas. The 15 acres adjacent to this on the southwest is owned by the University of Kansas Endowment Association, but is used by Mr. E. H. Chamney for the grazing of cattle. This portion is bounded on the west by a stone fence, beyond which lies a 14-acre field of natural prairie owned by Dr. C. D. Clark that is bordered on the extreme west by a narrow thicket of elm saplings.

The principal topographic feature of the area is an arm of Mount Oread, that rises some 80 feet above the surrounding countryside.

About 200 feet from the crest of the southwestern slope of the hill a 40-foot-wide diversion terrace directs run-off toward the two-acre reservoir that is the source of water for eight experimental fish ponds of the laboratory.

The predominant shrub-vegetation consists of Osage orange (*Maclura pomifera*), honey locust (*Gleditsia triacanthos*), and American elm (*Ulmus americana*). These saplings, ranging in height from 3 to 25 feet, grow in dense thickets as well as singly and in clumps of twos and threes. Larger trees of these same species grow along the crest of the hill, along the eastern and southeastern boundaries of the area, and along the stone fence separating University land from that owned by Dr. Clark. A dense growth of coralberry (*Symphoricarpos orbiculatus*) forms the understory just below the crest of the hill.

Isolated clumps of dogwood (*Cornus drummondi*) and hawthorn (*Crataegus mollis*) are scattered throughout the area. These species of shrubs grow densely along the stone fence. The isolated thicket on the Clark land is composed primarily of elm and boxelder (*Acer negundo*), but includes scattered clumps of dogwood, Osage orange, and honey locust. Poplars (*Populus deltoides*) are the only large trees in this area.

FIG. 1. Map of the study area near the University of Kansas Laboratory of Aquatic Biology. The dashed lines mark the approximate territorial

boundaries of the original nine pairs of Bell Vireos from May 1960 to early June 1960.

The open areas between the thickets are grown up in red top (*Triodia flava*), bluestem (*Andropogon scoparius*), Switchgrass (*Panicum virgatum*), Kentucky bluegrass (*Poa pratensis*), bush clover (*Lespedeza capitata*) and mullen (*Verbascum thapsus*). Shrubby vegetation occupies about 65 per cent of the total area, but in the Clark portion constitutes only about 35 per cent of the ground cover.

Considerations of Habitat

Nolan (1960:226), summarizing the available information on habitat preferences of the Bell Vireo, indicates that this species tolerates "a rather wide range of differences in cover." He pointed out that a significant factor in habitat selection by this species may be avoidance of the White-eyed Vireo (*V. griseus*) where the two species are sympatric.

In Douglas County where the Bell Vireo is the common species, the White-eyed Vireo reaches the western extent of its known breeding range in Kansas. At the Natural History Reservation of the University of Kansas, where both species breed, the Bell Vireo occurs in "brush thickets in open places" (Fitch, 1958:270) and the White-eyed Vireo occupies "brush thickets, scrubby woodland and woodland edge" (Fitch, *op. cit.*, 268). Along the Missouri River in extreme northeastern Kansas, Linsdale (1928:588-589) found the White-eyed Vireo "at the edge of the timber on the bluff, and in small clearings in the timber," while "the Bell Vireo was characteristic of the growths of willow thickets on newly formed sand bars." Elsewhere in northeastern Kansas I have found the Bell Vireo in shrubbery of varying density and often in habitat indistinguishable from that occupied by White-eyed Vireos at the Natural History Reservation. In the periphery of the region of sympatry the rarer species is confronted with a much higher population density of the common species and consequently might well be limited primarily to habitat less suitable for the common species.

This would seem to be the case in eastern Kansas, presuming that interspecific competition exists.

The Bell Vireo has followed the prairie peninsula into Indiana, aided by the development of land for agriculture. In nearby Kentucky where thousands of miles of forest edge are found, and where little brushy habitat of the type preferred by the Bell Vireo occurs, the White-eyed

Vireo is abundant whereas the Bell Vireo is unknown as a breeding bird (R. M. Mengel, personal communication).

In more central portions of the area of sympatry, nevertheless, the two species do occur within the same habitat (Ridgway, 1889:191; Bent, 1950:254) and occasionally within the same thicket (Ridgway, in Pitelka and Koestner, 1942:105); their morphological and behavioral differences, although slight, probably minimize interspecific conflict. The Bell Vireo and the Black-capped Vireo (*V.*

atricapillus) have been found nesting in the same tree in Oklahoma by Bunker (1910:72); the nest of the black-cap was situated centrally and that of the Bell Vireo peripherally in the tree. Bell Vireos invariably place their nests in the outer portions of trees and peripherally in thickets. This placement would further obviate interspecific conflict with the white-eye since its nests are placed centrally in the denser portions of a thicket.

A critical feature of the habitat preferred by the Bell Vireo is the presence of water. In far western Kansas this species is restricted to riparian growth along the more permanent waterways. This in itself is not adequate proof of the significance of water supply because thicket growth in that part of the state is found only along waterways. The 20

areas over the state that I have visited where Bell Vireos were present were closely associated with at least a semi-permanent source of water. Fifteen other areas indistinguishable from the 20 just mentioned, but lacking a permanent supply of water, also lacked Bell Vireos. Nevertheless areas in which Bell Vireos typically nest are decidedly less mesic than those frequented by White-eyed Vireos.

Once the Bell Vireo was probably more local in its distribution being restricted to thickets associated with permanent water. Clearing of woodland for agricultural and other use, and subsequent encroachment of second growth concomitant with the creation of man-made lakes and ponds, has greatly increased the available habitat for this bird. The preferred species of shrubs for nesting are reported (Bent, 1950:254) to be various wild plums (*Prunus sp.*). The widespread distribution and abundance of the exotic Osage orange has greatly augmented the supply of trees suitable for nesting.

SEASONAL MOVEMENT

Arrival in Spring

The subspecies of the Bell Vireo breeding in Kansas, *V. b. bellii*, winters regularly from Guerrero and the Isthmus of Tehuantepec south to Guatemala, El Salvador, and northern Nicaragua (A. O. U. Check-list, Fifth Edition, 1957:469-470). In the United States migrating birds are first recorded in early March (Cooke, 1909:119).

The Bell Vireo is a relatively slow migrator, moving primarily at night and covering little more than 20 miles at a time (Cooke, *op. cit.* 119). The average date of arrival, based on 27 records, for northeastern Kansas is May 8; the earliest record is April 22, 1925, from Manhattan, Riley County, Kansas (fig. 2-A).

In 1959 the first bird arrived at the study tract about May 5. No additional birds were heard singing until the third week of the month, in which eight new males were noted. As mentioned, in 1960 field work was begun in mid-April and the study area was traversed daily. No birds were detected until late afternoon of May 3, when one, presumably a male, was seen foraging.

Lawrence (1953:50) has reported that males of the Red-eyed Vireo precede females in the breeding area by as much as two weeks; the male Red-eyed Vireo forages but sings little in the pre-nesting period. The male Bell Vireo arrives first at the breeding area but precedes the female by only a few days. On the morning of May 4 the first male was singing from a number of perches while ranging over an area of seven acres. This area encompassed territories later occupied by three pairs, 2 (1960), 4 (1960), and 5 (1960). Late on the afternoon of May 4 the first courtship songs were heard and the first male was seen with a mate at 6:20 p.m.

Eight additional males arrived from May 6 through May 18. A tenth male was discovered in the vicinity of territory 9 (1960) on June 18, 1960.

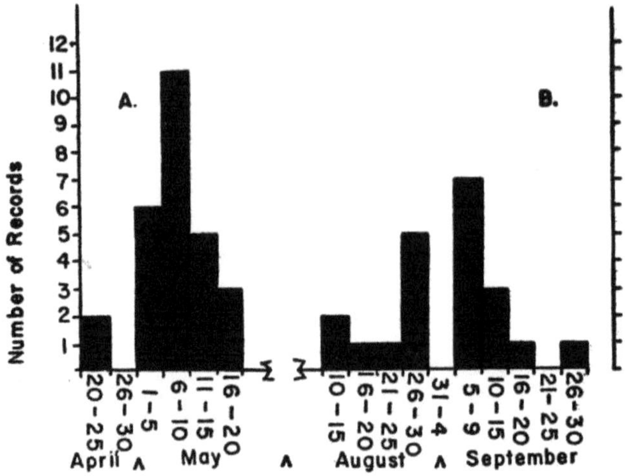

FIG. 2. Seasonal movement as indicated by the curve for spring arrival (A), based on the earliest dates for 27 years, and the curve for autumn departure (B), based on the latest dates for 21 years in northeastern Kansas.

Fall Departure

The average date of departure for 21 years in northeastern Kansas is September 3 (fig. 2-B). The earliest date is August 14 from Concordia, Cloud County, Kansas (Porter, unpublished field notes). The latest date is September 27 (Bent, 1950:262) from Onaga, Pottawatomie County, Kansas. In 1959 the last vireo was seen at the study tract on September 14. The birds do not all depart at the

same time. On September 1 there were still five singing males in the study area; by September 10 there were three and on September 13, only one.

GENERAL BEHAVIOR

Flight

In "straight-away" flight the Bell Vireo undulates slightly. In a typical flight several rapid, but shallow, wing beats precede a fixed-wing glide of from 1 to 15 feet in length. Because the wings are extended horizontally during the glide, the bird does not move distinctly above or below the plane of flight. The White-eyed Vireo

generally appears to be slower and more lethargic in flight than the Bell Vireo. In the breeding season most flights of the Bell Vireo are no longer than a few feet between adjacent shrubs and trees, but occasional sustained flights are as long as 300 feet. The birds fly as low as 2 feet above ground, but have often been observed as high as 70 feet above the ground.

In courtship and protracted territorial disputes, where chase between sexual partners or a pair of antagonists occurs, looping flights are observed. The wings are beaten as the birds climb and many aerial maneuvers are performed in the course of the glide.

Foraging and Food Habits

The Bell Vireo has been characterized as a thicket forager (Hamilton, 1958:311; Pitelka and Koestner, 1942:104), but in my experience it is not restricted to low level strata; birds forage from ground level upward, both in thickets and isolated trees ranging in height from 3 feet to 65 feet. The tendency to forage at higher levels is in part dictated by the presence of tall trees within the various territories.

Territories 1 through 7 (1960) contained from three to ten trees surpassing 40 feet in height. They grew singly or in small groves.

The Bell Vireos foraged fully 20 per cent of the time in these trees.

Food was sought throughout the leaf canopy.

Behavior in foraging in larger trees followed a routine pattern.

Typically a pair alighted in a tree at a height of 15 feet. Then the female hopped to a perch a foot above the one upon which she landed. The male succeeded her to the perch she had previously occupied. The pair in effect spiraled around some large, essentially upright, branch, in foraging. The birds usually reached higher perches in this manner rather than by flying upward 10 to 15 feet to them. This manner of progression within a tree is reminiscent of a similar habit of the *Cyanocitta* jays. Presumably, the habit of the Bell Vireo of foraging in higher strata is facilitated by the absence of other species of arboreal foraging vireos.

Chapin (1925:25) found the Bell Vireo to be more insectivorous in its food habits than any other North American vireo. He found 99.3 per cent of all food contained in 52 stomachs to be of animal origin. Only three times have I seen a Bell Vireo take food of vegetable origin. On September 9, 10, and 14, 1959, I noted a male eating wild cherries over a period of 65 minutes of observation.

Chapin (1925:27) noted that beginning in July vegetable matter represented 1.57 per cent of the bird's subsistence, and thereafter slightly more until fall migration.

Animal food, consisting primarily of insects and spiders, is actively sought along branches and under leaves. Often a foraging bird will leap to the underside of a branch and hover, mothlike, beneath a cluster of leaves while extracting some insect. Some individuals hung upside down on small branches, paridlike, while foraging.

Lawrence (1953:710), and Southern (1958:201) have recorded similar behavior of the Red-eyed Vireo. Occasionally, I have seen a Bell Vireo fly from a perch and capture an insect in the manner of a flycatcher. The birds do not appear to be adept at this type of food-getting. Nolan (1960:242) mentions Bell Vireos holding hard-bodied insects by means of their feet while breaking the exoskeleton with the beak to obtain the soft parts. Southern (1958:201) recorded a female Red-eyed Vireo foraging on the ground; I have seen a Bell Vireo on the ground but once, and it was gathering nesting material.

Bathing

On May 14, 1960, in a rill that empties into the northeastern edge of the reservoir a female flew down from a perch six inches above the surface, barely dipped into the water, flew to a perch 12 inches above the water, violently shook her ruffled body feathers, quivered her wings, and rapidly flicked her fanned tail. The entire procedure was repeated three times in five minutes. She was accompanied by a singing male that did not bathe.

Nolan (1960:241) reports a male Bell Vireo bathing by rubbing against leaves wet with dew; he notes that the White-eyed Vireo bathes in a similar manner. Southern (1958:201) twice observed Red-eyed Vireos bathing in water that dropped from wet leaves.

In my study area in 1960, only territories 7, 8, 9, and 10 were not immediately adjacent to permanent water. The pairs of Bell Vireos in those territories presumably had to reply on wet vegetation for bathing.

VOCALIZATIONS

The male Bell Vireo begins to sing regularly soon after its arrival in spring. Some daily singing continues following the cessation of breeding activities until departure of the species in late summer or early fall. The highest sustained rate of song occurs on the first and second days of nest building. Because careful records of meteorological data were not kept, I cannot significantly correlate rates of song and specific temperatures and other weather conditions. Frequency of song was reduced when the temperature rose above 90deg F., as it did on many days in June, 1960. Nice (1929:17) mentions a similar decrease in singing when the temperature exceeded 85deg F.

Passerine birds typically sing at a high rate throughout courtship and nestbuilding, but at a markedly lower rate thereafter. Most vireos are atypical in this respect. In the study area in 1960 Bell Vireos sang more often than Robins, Mockingbirds, Field Sparrows, Brown Thrashers, Catbirds, and Doves breeding in the same habitat, about as often as the Meadow Larks in the adjacent fields, and less often than Painted Buntings.

The Bell Vireo seems to sing less often in the undisturbed state than when aware of the presence of an observer. Observations from my car, at a site approximately equidistant from territories 1 (1960), 2

(1960), 4 (1960), and 6 (1960) indicate that the rate of song during incubation is decidedly less when no disturbing influence is present.

Normally, in this period, song aids in maintaining contact between the members of a pair, serving to locate the male as he forages. Mumford (1952:230) noted that the males often came out to meet him as he entered their territories, singing as they approached. The male typically continues to sing for some time after the intruder has departed. Here the song acquires the additional functions of alerting the female to danger and threatening the trespasser. Even after allowance is made for this reaction to disturbance, Bell Vireos sing more often than most of their nesting associates, and, on a seasonal basis, they are vocal for a much longer time.

Singing Postures

In the normal singing posture the body of the Bell Vireo is maintained at an angle of 35deg to the horizontal. Occasionally, during nest building, I have observed the body held at angles as severe as 80deg from the horizontal.

The head of the White-eyed Vireo is distinctly bobbed up and down, two or three times, during the utterance of a song phrase.

A bob involves a deliberate withdrawal of the head towards the body and subsequent sharp, almost vertical, extension of the neck.

The head of the Bell Vireo does not bob, although it vibrates as the song is delivered.

Flight Song

The Bell Vireo does not have a distinctive flight song; in fact, it rarely sings or calls while in flight. Nolan (1960:240) has recorded a male singing the normal song while in flight. Sharp scold-notes are uttered in mid-air when a bird is agitated or actually attacking an enemy. These notes and songs recorded by Nolan hardly qualify as flight song, for this term implies use of a distinctive vocalization not uttered in other circumstances.

Daily Frequency of Song

In the morning, Bell Vireos usually began singing a few minutes before sunrise. Their songs were invariably preceded in the study area by those of Western Kingbirds, Robins, Mourning Doves, Mockingbirds, Cardinals and Meadow Larks. Bell Vireos sang relatively little after 6:30 p.m., even on the longest days of the year.

The latest daytime singing that was recorded was seven songs at 7:18 p.m. on June 20, 1960. A Cardinal in the vicinity sang for a full hour after this.

Types of Vocalizations

Six vocalizations were readily distinguishable in the field. These are divisible into songs and call notes.

1. Primary song. It has been described by Pitelka and Koestner (1942:103) as an "irregular series of harsh and sharp, but slurred notes preceded by a few distinct notes of the same quality and ending with a decided ascending or descending note of similar harshness." The

terminal note may also be somewhat abbreviated and intermediate between an ascending or descending note. The song is sometimes delivered as a couplet that consists of a phrase ending on a descending note. This delivery is typical of incubation and later renesting. During early season activities, the bird utters a phrase ending on the descending note as many as 15 times before a phrase ending on an ascending note is heard.

A sonagram of a single phrase, one of several recorded on May 9, 1960 (the third day of building of nest 1-b 1960), consists of 10 notes, the first of which is distinct. The remaining notes are slurred. This phrase is 1.4 seconds in length.

Songs are delivered most rapidly in the course of territorial disputes and defense. The song is loudest in times of nestbuilding and periods of aggressive behavior. At these times, on clear, calm days, the songs are audible 100 yards away. Singing in the nestling period and post-breeding season is audible at distances of no more than 50 feet; such notes have been termed "whisper songs." Table 1 summarizes singing rates at different periods of the nesting cycle in several situations and under various weather conditions.

Songs are of equal frequency in the immediate vicinity of the nest and elsewhere in the territory. Nice (1929:17) also found this to be true. Perches can be almost at ground level or as high as 60

feet. Forty per cent of my data on song concern singing at heights of more than 20 feet. As indicated in foraging, the lack of competition from aboreal species of vireos presumably contributes to the use of higher perches by Bell Vireos.

No female song was recorded in 1959, but on May 26, 1960, a female was heard to sing once. She appeared at nest 1-f (1960) shortly after the male arrived. Unlike him, she did not participate in building, but seemed to be inspecting the nest. After 30 seconds she sang once--a low garbled phrase--and also scolded once. After this she left. In the meantime the continuously singing male moved two feet away from the nest, then back to it and resumed construction.

The song of the female signaled to the male her departure.

Pitelka and Koestner (1942:103) heard a female sing twice after she replaced the male on the nest. Females of three other species of vireos, the Black-capped Vireo, *V. atricapillus* (Lloyd, 1887:295), the Philadelphia Vireo, *V. philadelphicus* (Lewis, 1921:33), and the Latimer Vireo, *V. latimeri* (Spaulding *in* Pitelka and Koestner, 1942:103) have been heard singing.

Lewis and Spaulding also suggest that the song of the female functions as a signal prior to exchange at the nest.

The primary song identifies the singer as a male Bell Vireo. It aids in securing a mate and in warning potential adversaries; also, the song is a signal in certain situations and serves to locate the male.

TABLE 1. REPRESENTATIVE SINGING RATES OF BREEDING BELL VIREOS. ALL RATES WERE AT AIR TEMPERATURES LESS THAN 86DEG F. EACH INSTANCE REPRESENTS APPROXIMATELY 30 MINUTES OF OBSERVATION.

Circumstance	Instances	Average rate per minute
Attraction of mate	2	6.3
Territorial dispute	5	12.8
Nestbuilding	6	7.0
Egglaying	1	3.0
Incubation	6	3.9
Exchange of partners in the incubation period	1	4.0[A]
Foraging	2	2.2
"Morning" song	1	28.6[A]

"Evening" song

1

1.9[A]

Overall average rate per minute 6.3

[A] Not sustained; data representative of periods less than 5 minutes in length.

2. Courtship song. It is here termed the "congested" song and is comparable to the adult "run-on" song mentioned by Nolan (1960:240).

The congested song is a squeaky version of the primary song and is given when birds are engaged in pair-formation, nestbuilding, and egglaying. The delivery is rapid and the sound can be likened to that made by rapidly scraping a bow across a taut violin string.

Nolan (*in* Mumford, 1952:230) is probably speaking of this song when he describes a "tuneless" song that "had a jerky, sputtering quality that characterizes part of the song of the Ruby-crowned Kinglet (*Regulus calendula*)." More recently (1960:240) he applies the adjectives "twanging," "Bobolink-like," "bubbling," "jerky," and "squeaky." This song is often blended with the primary song and is audible for 75 feet.

A specialized version of the congested song is associated with pre-and post-copulatory display but differs from the typical squeaky performance in terminating in two ascending notes reminiscent of the ascending phrase of the primary song.

3. Distress call. It was heard only once, when a captured bird was being freed from a net. When the bird was almost disentangled it uttered 10 high-pitched, plaintive notes. The quality of the notes suggested a relationship to the song phrase rather than to other types of vocalization. A nesting pair of Bell Vireos, 10 feet away, became extremely excited when they heard the distress notes. They "scolded" vigorously and flew around my head at a distance of six feet.

4. Alarm note. This is a specialized, three-note call of the male and was heard only from the onset of pair-formation through early nestbuilding. This whinnying, flickerlike call, phonetically *eh-eH-EH*, each succeeding note of which is louder than the one before, is given whenever the male is disturbed by an unfamiliar object. This call is generally succeeded by the *chee*, but occasionally blends into an extended "whinny," and is typically given from some perch affording an unobstructed view of the offending object.

The male stretches his neck and cocks his head, the wings and tail are not flicked or fanned, and no feather tracts are erected. The bird, nevertheless, flits nervously from perch to perch when uttering the call.

5. The *zip*. The male has a special "scold" note of his own that is heard when an intruder first approaches the nest. Phonetically it is *zip-zip-zip*. It is not so loud as the *chee*, and the delivery is more deliberate than that note. If the intruder remains near the nest, the *zip* is usually replaced by the *chee*.

6. The generalized call note or *chee*. The call notes associated with several situations are combined under this subheading since all can be rendered in English by the same phonetic equivalent--*chee*. The *chee* associated with nestbuilding is of moderate pitch and delivered deliberately at a rate of about 40 per minute. The feeding call of the adults is a soft slurred *chee*, while that of the nestlings has a mewing quality. In general, the *chee* utilized in signal situations consists of a few repetitions of the basic note emitted at a moderate pitch. The *chee* associated with hostile and courtship behavior is higher pitched and the delivery is much more rapid, approximately 200

per minute. Nolan (1960:240) reports a continuous rate of 25 per five seconds when an adult Bell Vireo is alarmed. The *chee* of extreme anxiety is a loud emphatic buzz, phonetically ZZ-ZZ-ZZ-ZZ.

TERRITORIALITY

The Bell Vireo exhibits "classic" passerine territoriality. Within a specific area, a pair of this species carries out pair-formation, courtship activities, copulation, nesting, rearing the young, and foraging. With the cessation of reproductive activities, a pair continues to restrict its other daily activities to the same general area.

Establishment of Territory

In early May the segment of the total suitable habitat within which a Bell Vireo restricts its activities is not rigidly defined and the first male of the season ranges over an area too large to be maintained permanently -- one that seems greatly to exceed the needs of breeding. Male 1 (1960), for instance, was first seen foraging over an area of approximately seven acres. With the influx of other males, portions of this large tract were usurped and the territory of the original male was gradually reduced to an area of little more than an acre.

In this initial period, a male becomes identified with a large area but is restricted to an area of nearly typical size by the encroachment of other males. Territorial disputes in this period often involve physical contact, as well as protracted sessions of high-intensity singing at rates exceeding three hundred song-phrases per hour.

Eventually the carrying capacity of the habitat is reached and no further partitioning occurs. The beginning of nestbuilding coincides with this relative stabilization of the territorial boundaries.

Through the remainder of the cycle of behavior associated with any one nest, all activity is that of the occupant pair within its territory.

Size of Territories

The nine original territories established in 1960 varied in size from 0.26 acre to 3.1 acres (Table 2). Fitch (1958:270) found the territories of several pairs of Bell Vireos at the University of Kansas Natural History Reservation to vary from 0.4 to 1 acre. Hensley (1950:243) estimated the size of the territory of a pair of Bell Vireos observed in Piatt County, Illinois, at 3.1 acres. Nolan (1960:227) records home ranges of 2 to 3 acres.

The pairs that he studied were sole occupants of fields several acres larger than the portions actually utilized. His description of the vegetation indicates that most of the second growth was not much taller than 7 feet. As indicated elsewhere, the second-growth in my tract averaged 15 feet tall. The smaller average size of territory (1.25 acres) that I found is probably a function both of this greater vertical range of available foraging area and the much higher gross density of birds (40 pairs per 100 acres).

Permanence of Territories

Most pairs remain in their original territories throughout the summer, although some shift certain territorial boundaries. In 1960 pairs 2 and 6, in the course of selecting a site for a replacement nest, annexed adjacent areas previously occupied by other pairs. Pair 2 relocated in a space that originally included territories 1 and 4, and pair 6 built a nest in an area formerly occupied by pair 7. Males 1 and 4 were sacrificed for specimens and pair 7 probably was destroyed by a predator. Owing to the presence of a nest, the annexed area becomes the focal point of the activities of a pair, but the original area is regularly visited and may be returned to in a later renesting.

TABLE 2. SIZE OF THE NINE ORIGINAL TERRITORIES OCCUPIED IN 1960.

Territory	Date first occupied	Dimensions
1.	May 3, 1960	1.6 acres
2.	May 5, 1960	0.6 acre
3.		

May 7, 1960
0.26 acre
4.
May 11, 1960
1.03 acres
5.
May 12, 1960
2.07 acres
6.
May 14, 1960
3.1 acres
7.
May 13, 1960
1.7 acres
8.
May 14, 1960
0.46 acre
9.
May 14, 1960
0.4 acre
Average 1.25 acres

Maintenance of Territory

Except in the early stages of nesting, territory is maintained primarily by song. In the period of incubation a male regularly patrols his territory between sessions of sitting on the eggs. He sings several songs from each of several perches. A male follows a predictable path, rarely traveling more than 150 feet from the nest. Incipient patrolling is seen early in the breeding season when territorial boundaries are in a state of flux.

The male White-eyed Vireo travels a semi-predictable route, as does the Solitary Vireo (R. F. Johnston, MS). According to Lawrence (1953:50), the male Red-eyed Vireo has a distinct singing area completely divorced from the nest area dominated by the female.

Southern (1958:109), working with this same species in Michigan, did not recognize separate areas, but found that the male wandered randomly over the territory.

In a species so highly active as the Bell Vireo, the degrees of hostile action associated with an encounter overlap in such a fashion that no clearcut distinction can be drawn among the various displays. Nevertheless, certain generalized patterns are characteristic of all situations in which members of this species are in a state of anxiety. The threat displays described in the succeeding paragraphs may all be utilized within as little as two minutes; mutual agonism may be terminated at any stage by concerted attack of the dominant bird.

1. Vocal threat. When an intruder is discovered the resident male markedly increases his rate of singing. The alarm note, *eh-eH-EH,* is the first call uttered during the nestbuilding and egglaying periods.

2. Head-forward threat. If the intruder does not flee, the resident male adopts a specific threat posture. The head and neck

are extended. The feathers of the crown are erected, but those of the body are sleeked. The bird crouches slightly and the tail is flicked laterally, but not fanned. The intensity of the singing increases and is supplemented by scolding, also delivered at a rapid rate. The intruder normally retreats at this juncture.

3. Wing-flicking and submaximal tail-fanning. If the interloper remains, the anxiety of the resident male increases. He slightly depresses the tail and, at the same time, rapidly fans and closes it.

The tail is only partially fanned. The wings are held slightly away from the body and rapidly flicked above the back. This flicking should not be confused with quivering of the wings associated with begging and other solicitory postures. Song is now almost completely replaced by high-intensity scolding. Associated with this

high degree of anxiety are displacement behaviorisms, including bill-wiping, reversal of direction on a single perch, and a nervous hopping from one perch to another.

4. Ruffling and maximum tail-fanning. This display is most often seen in conjunction with the harassment of predators, but occasionally it is observed in territorial disputes occurring at the boundary of adjacent territories where neither male is strictly dominant and in which there is much vacillation prior to attack.

The feathers of the abdomen are ruffled. The term "ruffled" pertains to a full erection of the feathers, giving a ragged appearance to the body outline (Morris, 1956:80). Ruffling of the abdominal feathers emphasizes their yellow color and seemingly heightens the intimidatory effect. The tail is fully fanned, and so maintained, for a few seconds at a time; it is held at a 45deg angle to the body.

The scold becomes an extremely intense, stacatto buzz, ZZ-ZZ-ZZ-ZZ.

5. Supplanting attack. The attack directed against a trespassing male is initiated as a lunge that results in a collision with the opponent in mid-air or on his perch. The bird attacked is struck by his adversary's open beak or body.

Hinde (1952:71-72) indicates four courses of action followed by a Great Tit (*Parus major*) when attacked under similar circumstances.

"(a) It flies away: The attacker usually flies after it and a chase ensues. (b) It shifts its perch a few inches: the attacker lands in its place, and both usually show head-up postures. (c) It remains where it is, but adopts a head-up posture: the attacker usually then shows upright flight. (d) It may fly up and meet the attacker in mid-air: in that case an actual combat may result, or both combatants may show upright flight."

Head-up posturing and upright flight are not presently recognized components of the behavior of the Bell Vireo. The behavior of the attacked Bell Vireo is similar to that described in (a), (b), and (d) above, and is clearly dictated by the proximity of his own "home base."

Eleven disputes among occupants of adjacent territories were witnessed between May 6 and June 3, 1960, in which some or all of the described threat displays were manifest (Table 3). In each instance, patrolling males were gradually attracted to each other.

As they approached, their rates of song increased from an average of six repetitions per minute to 15 per minute. Eight of the disputes involved physical combat.

On May 6, 1960, when male 2 (1960) was in the process of usurping an eastern segment of the original territory of male 1 (1960), a violent, protracted dispute was observed. By this date male 1

(1960) had obtained a mate and had begun construction of nest 1-a (1960); male 2 (1960) had not yet acquired a mate. At first the two males were singing vigorously, from one to 10 feet apart.

Female 1 (1960) followed her mate closely and scolded, at the same time partially fanning her tail. In the course of vocal dueling the males had traveled to within 50 feet of nest 1-a (1960), when male 1

(1960) suddenly lunged at 2 (1960). The males plunged to the ground, locking bills and clutching at each other with their feet as they fell. As soon as they touched the ground they separated.

Male 2 flew east with male 1 in pursuit. This conflict lasted three minutes.

Additional physical combat was witnessed several minutes later.

This again involved striking with the bill, wings and feet. A high pitched squeaky *chee* was uttered by both combatants. The female scolded from a nearby perch. Upon separating, the males engaged in a wild, looping flight. At about 350 feet from nest 1-a (1960), the chase abruptly ended. For ten minutes thereafter, both males sang at a high rate from perches about 10 feet apart. This terminated the physical combat, but three additional protracted, vocal duels occurred in the remainder of the morning.

TABLE 3. INTRASPECIFIC DISPUTES IN MAINTENANCE OF TERRITORY.

Behavior	Number of conflicts	Vocal dueling	Combat	Average length of disputes
Prenesting	3	3	2	6 min. 40 sec.
Building	8	8	6	3 min. 8 sec.
Incubation	1[B]	1	...	20 min.
Totals	**12**	**12**	**8**	**5 min. 30 sec.**

[B] Directed against a stuffed Bell Vireo.

Probably as a direct result of these conflicts, a neutral zone approximately 300 feet wide developed between the two territories. By May 14 this intervening area was occupied by male 4 (1960). By this date both 1 (1960) and 2 (1960) were involved in nestbuilding and 4

(1960) was not challenged for several days.

Maximum tail-fanning prior to attack also appears as an element of aggressive behavior in White-eyed Vireos. A brief skirmish between a male of this species and a small, greenish passerine was observed at the Natural History Reservation on May 25, 1960. The White-eyed Vireo was singing from a perch 30 feet high in a dead elm, when the unidentified passerine landed 10 feet distant. The white-eye ceased regular song and uttered several catbirdlike calls, and at the same time slightly depressed and fully fanned the tail.

After 10 seconds, the white-eye lunged at the intruder, striking it in mid-air. A brief looping flight ensued through the branches of the elm before the intruder was able effectively to retreat.

Aggressive Behavior of the Female

The female Bell Vireo is concerned primarily with the defense of the nest and the young and she rarely assists the male in defense of distant parts of the territory. She employs the same threat displays as the male.

Interspecific Relationships

A number of meetings between Bell Vireos and other species were observed in the course of the study (Table 4). Resident pairs of this species exhibited different degrees of tolerance toward other species. Many birds, including Cardinals, Field Sparrows, Painted Buntings and Mourning Doves were ignored completely. Chickadees evoked responses characterized by slight increase in song and some anxiety; this was perhaps owing to similarity in size, motion and call notes. Warblers, when met with, were invariably chased.

They may be momentarily mistaken for rival vireos.

TABLE 4. INTERSPECIFIC CONFLICT OBSERVED IN 1959 AND 1960.

Species

Number of conflicts

Phase of breeding cycle

Behavior of Bell Vireos

HFT[C]

S

TF

A

Coccyzus americanus

1

Nestling period

x

Cyanocitta cristata

3[D]

Nestling and incubation period

x

x

x

x

Parus atricapillus

1

Prenesting

x

Molothrus ater

1

Nestling period

x

x

Dendroica petechia

1

Prenesting

x

x

Geothlypis trichas

1

Nestbuilding

x

x

Pituophis catenifer[E]

1

Post-fledging

x

x

[C] HFT = head-forward threat; S = scolding; TF = tail-fanning; A = attack.

[D] Includes attack against a dummy Blue Jay.

[E] The Bull Snake is here included because the vireos directed typical aggressive displays towards it.

Blue Jays were vigorously attacked, especially late in incubation and throughout the nestling period of the Bell Vireo. I did not see a jay struck, but a vireo would circle one closely as it perched and pursue it when it flew, following as far as 100 yards beyond territorial bounds. The buzz, ZZ-ZZ-ZZ-ZZ, was uttered in conjunction

with this harassment.

A stuffed jay placed eight feet from a nest elicited threat display and displacement behavior from the owners of the nest, but no attack. Incubation had just begun at this nest. A dummy Bell Vireo placed close to another nest only momentarily disturbed the male, and the female completely ignored it. Incubation had also recently begun at this nest. At this same general stage, moreover, nesting pairs showed little inclination to harass me.

Discussion

Hinde (1956:341-342) indicates that territory has been defined in a number of ways by many workers. All of the definitions involve modification of Howard's classic "defended area." Pitelka (1959:253) has reacted against this behaviorally-oriented concept. He thinks that the concept of territory should be based on exclusive use of an area by its occupants, and not so much the defense by which they maintain it.

Methods of treating territoriality in the Bell Vireo seemingly incorporate features of both schools of thought. The area used exclusively for all biological needs by a single pair of Bell Vireos is vigorously defended both physically and vocally early in the breeding season and vocally as the season progresses.

In the period of territorial establishment a relatively large area is actively defended. The building of a nest establishes a focal point of activity in a somewhat more restricted area than that originally occupied. After the success or failure of a nest, a new site is selected to which the focal point of activity is shifted. If suitable habitat adjacent to the extant territory is unoccupied by other Bell Vireos the unoccupied area may be annexed in the course of searching for a new site. Such annexation occurs only when pairs formerly occupying adjacent suitable habitat disappear from this territory;

possibly the size of the territory of any one pair is dictated by the density of population of the species as well as by the presence of suitable habitat. This may not always be true as indicated by Kliujver (1951:40), who in studying the Great Tit, found no appreciable difference in the size of territory in two different habitats even though there was a marked difference in population density of the birds.

Fluctuation of territorial boundaries is not uncommon in passerines, especially when no rivals exist to contest movement. Hinde (1956:351) indicates that fluctuations in size of territory are to be expected although the territories of different species of birds have different mean sizes.

Once nesting activities commence there is a marked reduction in the amount of territory utilized and a distinct decrease in the aggressive tendencies of the male; it would seem that energy previously utilized in regular fighting is rechanneled for nestbuilding, incubation and care of the young. Further, contraction of the area of activity obviates high-intensity territorial defense, as adjacent males, even in regions of high population density, are isolated from one another by an area no longer regularly traversed.

With cessation of breeding activities physiological mechanisms governing maintenance of territory seemingly are no longer active and yet the pairs of Bell Vireos remain within a restricted area which they alone use. Earlier definitions of territory as a "defended area"

do not adequately cover such situations and yet from the standpoint of Pitelka the area still retains the characteristics of true territory.

In fact, territory as defined by Pitelka is clearly manifest at this time. Whether the birds remain in an area through "force of habit" is of little consequence.

I have retained the term "territory" in preference to the term "home range" used by Nolan (1960:227). His failure to observe territorial defense is responsible for his terminology, although it is readily understandable that such defense would be lacking in a population of relatively low density in which pairs were isolated from one another by areas of unfavorable habitat. This isolation in itself would tend to preclude territorial conflict but territories were, in fact, maintained.

The marked similarity in the essential features of aggressive behavior in North American vireos attests to their close relationship.

Flicking and fanning of the tail are distinct components of the hostile behavior of the Bell Vireo, White-eyed Vireo, Red-eyed Vireo (Lawrence, 1953:69), and the Black-whiskered Vireo (*Vireo altiloquus*; Bent, 1950:319), and, presumably, of the remaining species of the genus. The occurrence of these same displays as intrinsic behavioral elements of interspecific hostility suggests a common derivation.

Moynihan (1955:256) indicates that all intraspecific hostile displays, and probably most interspecific hostile displays, evolved originally as social signals having the same general function. Further, Hinde (1956:344) points out that there is a fundamental similarity in the motor patterns used in fighting in different contexts, including both interspecific and intraspecific fighting.

COURTSHIP BEHAVIOR

The precise mechanism of pair-formation in the Bell Vireo is not known. My experience has been to find a male one day and then one or two days later to discover that it has a mate. Lawrence (1953:53), tells of a male Red-eyed Vireo singling out a female from a flock of migrants passing through his territory and violently driving her to the ground. Shortly after this attack the pair was seen searching for a nest site. But such an incident has not been reported for other vireos, nor have I witnessed such behavior myself.

Early courtship activities of the Bell Vireo are characteristically violent affairs, with the male directing strong aggressive attacks toward the female. Rapid, looping flights through the thickets occur, the female leading the male. Occasionally he deliberately collides with her in mid-air, but the pair quickly separate. This violent sexual chasing is manifest prior to the inception of nestbuilding.

With commencement of this activity, sexual chases through the territory subside.

Absence of sexual dimorphism in the Bell Vireo obviously suggests that behavioral criteria are used by the birds in sex-recognition. The lack of aggression by the female upon initial aggression by the male is an essential component of recognition of sex; she is clearly subordinate.

Such subordination is also the significant feature of continued sex-recognition. Courtship display by a resident male, directed toward a stuffed male and a wounded male which sat motionless, supports the contention that a subordinate or submissive attitude of the female is a key factor in sex-determination.

Nestbuilding and courtship are intimately associated in this species. The male constructs the suspension apparatus of the nest, the completion of which coincides with the assumption of nestbuilding activity by the female. Roles of the sexes in nestbuilding are described in the section on nestbuilding. The male frequently interrupts construction to court the female. This, in combination with perpetual song as he works, serves to strengthen the pair-bond and stimulate nestbuilding tendencies of the female.

It is doubtful that any attempts at copulation are successful up to this time. The female is singularly unresponsive to the advances of the male; a female retreats before most violent attacks and is seemingly oblivious to less vigorous behavior. After the female assumes the responsibility of building, the tempo of courtship activities increases.

The female becomes increasingly more receptive and her work is often interrupted by advances of the male. Copulation occurs frequently from about the third day of nestbuilding through the first day of egglaying, a period of four to six days. Male displays and vocalizations associated with courtship continue through the fourth or fifth day of incubation.

Displays and Postures

The principal courtship displays and postures that were seen throughout the nestbuilding phase are as follows:

1. Greeting ceremonies. Both birds are crouched from one to five inches apart. The feathers on one (the male?) are sleeked, and on the other are fluffed. Fluffing (Morris, 1956:80) denotes partial erection of the body feathers producing a rounded, unbroken body line and is not to be confused with ruffling, mentioned in the sections pertaining to territoriality and pre-and post-copulatory display.

Fluffing is generally considered to be an appeasement display and it is seen in a variety of situations involving a dominant-subordinate relationship. Both birds flick wings and tails rapidly and reverse directions on their perches frequently. A low, rapid *chee* is uttered during this performance. This ceremony is repeated often in the first three days of nestbuilding, but less frequently thereafter. It usually occurs after building by one or both partners and prior to another trip in search of nesting material. It lasts from 10 to 50

seconds and is not immediately followed by any additional courtship activities. Nolan (1960:228-229) observed mutual displays between periods of violent sexual chase that suggest that the greeting ceremonies that I have described are an integral part of pair-formation as well as a component of continued maintenance of the bond.

2. "Pouncing." The female rapidly quarter-fans and partially depresses her tail. She utters a high pitched scold (*chee*). The male, from a perch within two feet of the female, fans the tail fully and depresses it vertically, and, with mouth open, lunges at the female; or, with similar tail mannerisms, the abdominal feathers ruffled, the wings held horizontally, and the primaries spread, he sways from side to side from four to six times, and then lunges at the female. The male is silent when he pounces; the *chee* or the courtship song is emitted when swaying precedes pouncing. The male strikes the female with his breast or with

his open beak. The female rarely flees although she is usually displaced several inches along the branch upon which she is sitting. However, the female may fly several inches to a new perch. The failure of the female to adopt a solicitation posture presumably indicates sexual unreadiness.

Instances of the male deliberately colliding with the female as she flies in the course of gathering nesting material are probably analogous to pouncing. In none of the above situations are females observed to fight back in any way. Nice (1943:174) believed pouncing to be analogous to sexual chasing found in such species as the Red-winged Blackbird. In the Song Sparrow, pouncing is observed most often in the first and second days of nestbuilding.

3. "Leap-flutter." The male, in the course of displaying with the tail fanned before the female, suddenly leaps eight inches to ten inches vertically and flutters in mid-air several seconds, before dropping to the original perch. This display occurs in full view of the female. It is often associated with pouncing and is also seen prior to copulation. In the latter instance it is probably pragmatically functional, for it permits the male to orient above the female before dropping to her back to copulate. No vocalization is uttered during the leap-flutter.

FIG. 3. A single male Bell Vireo in the pre-copulatory display. Note the ruffled dorsal and ventral body feathers.

The male on the left has reached the zenith of a single swing. The male on the right has nearly reached the low point of a swing.

4. Pre-copulatory display (Fig. 3). The male faces the female.

The tail is fanned fully and depressed at a sharp vertical angle to the body. Body feathers, both dorsal and ventral, are ruffled, almost tripling the apparent volume of the thorax. The head is withdrawn and slightly thrown back. Feathers of the head are not erected.

The mouth is opened wide. The legs are slightly flexed and the body is swayed laterally. Horizontally, the head and body traverse an arc of about 100deg; vertically, they traverse an arc slightly less than 180deg. At the low point of any one swing, the delivery of the courtship song begins. At the termination of the swing the two normal, ascending notes are emitted. This performance may last as long as three minutes.

The pre-copulatory display of the male elicits receptive behavior in the female. She crouches in a solicitous manner, with the body feathers fluffed and the tail raised slightly, and utters a muted *chee*.

5. Copulation. The male abruptly terminates his swaying display with a leap-flutter that positions him above the female's back. He then descends and copulation occurs. The male continues to flutter his wings to maintain balance throughout the two seconds of cloacal contact. Following an unsuccessful copulation on June 23, 1960, displacement preening and bill wiping were performed by both sexes.

6. Post-copulatory display. On June 25, 1960, after a second attempt at copulation with a stuffed bird in which semen was actually deposited on the dummy's back, male 10 (1960) performed a swaying display. In this instance, however, instead of addressing the dummy from the front, the male alighted one inch to the right of the stuffed bird. When swaying to the left (toward the dummy) the head of the displaying male actually passed above the neck of the stuffed bird. This ritualized behavior could conceivably be derived from hetero-preening.

Discussion

Within the scope of my research it was difficult to detect the over-all sequence of epigamic displays that result in synchronization of the physiological states of the sexes throughout the period of courtship. Possibly all displays, except the post-copulatory one, occur in no particular order in the courtship period. However, each ritualized display seemingly strengthens the pair-bond.

Swaying has been recorded in a variety of situations of a sexual and semi-sexual nature for the Solitary Vireo (*V. solitarius*; Townsend, 1920:158) and the Red-eyed Vireo (Tyler, 1912:230; Bent,

1950:342). In every instance the body feathers of the swaying birds were sleeked. Courtship behavior in any species of North American vireo seems closely to resemble that of any other; pairing and nestbuilding of a female *V. solitarius* and a male *V. flavifrons* as reported by Hauser (1959:383) support the idea of close resemblance.

A marked similarity will be detected between certain basic elements of aggressive and epigamic displays. These basic elements

are wing-and tail-flicking, tail-fanning, and high-intensity delivery of the *chee*. Pouncing and supplanting attacks are essentially similar.

Such similarities suggest either a common origin for certain aggressive and epigamic displays or the derivation of one from the other.

High-intensity *cheeing* is obviously a function of excitement, whether in conjunction with hostility or sexual behavior. According to Andrew (1956:179), flicking of wing and tail in passerines are intention movements of flight. These actions have been emancipated from incomplete take-offs and incorporated in ritualized courtship and agonistic behavior. In incipient courtship behavior the male is governed by three conflicting tendencies; to flee, to attack, or to behave sexually before his mate (Tinbergen and Hinde, 1958:256). When pairing, Bell Vireos interrupt sexual chase with "greeting ceremonies," the male's tendency to attack and the female's tendency to flee are momentarily reduced, and the forming bond is strengthened.

Thus, the intention movements become an integral part of courtship.

In situations where attacking and fleeing are the two conflicting tendencies, wing-flicking and tail-flicking are incorporated into threat display, but do not lose all of their original function, for they facilitate attack. Tail-fanning, as a display element, increases the awesome aspect of the threatening bird and in courtship presumably makes the sexes more attractive to one another.

Courtship feeding has not been recorded for the Bell Vireo. In general, it is unknown in North American vireos, with the exception of the red-eye (Lawrence, 1953:53). It would serve no "practical"

purpose in the Bell Vireo since the male regularly relieves the female during incubation, thus allowing her ample opportunity to forage. In the Red-eyed Vireo, only the female regularly incubates, and courtship feeding is definitely functional. Nolan (1960:228) described a brief pecking or pulling with their bills between pairing birds. This may be incipient "symbolic" courtship feeding, or perhaps mutual preening.

SELECTION OF NEST-SITE AND NESTBUILDING

As far as can be determined, the nest-site is selected by the female. Typically, the pair makes short, low-level flights from tree to tree with the female invariably in the lead. The birds usually forage within each tree; the female interrupts this activity to inspect small forks of low, pendant branches and the male occasionally pauses to sing. The singing is loud but not particularly regular, as it is later when the male accompanies the female during actual nestbuilding. Method of selection of site resembles that described by Lawrence (1953:53) for the Red-eyed Vireo.

Nests are suspended from lateral or terminal forks about 27 inches high in bushes and small trees that, in the study area, averaged 11 feet, four inches in height (Table 5). The height above ground of the nests does not vary appreciably as the season progresses as is the case with nests of Red-eyed Vireos, for which Lawrence (1953:54) noted that late nests were placed higher than those built earlier in the season.

Most nests are so situated that they are protected and concealed by the dense foliage of trees. Where nests are placed in low bushes, as coralberry or dogwood, the bush is invariably overhung by the foliage of a much taller shrub or tree.

The nest tree or shrub was in every instance situated at the edge of a thicket or isolated from adjacent trees by several feet. Preference for open situations is characteristic of the species. In contrast, the nest of the White-eyed Vireo (Bent, 1950:229) is placed toward the center of thickets.

In the choice of sites in the study area, the Bell Vireos were almost unopposed by other avian species, owing to the size of the fork utilized and the fact that the nests are located peripherally, rather than centrally, in the bush or tree. This lack of competition for a nest-site provides a Bell Vireo with an ample supply of nest-sites within any one territory.

TABLE 5. NEST-SITES UTILIZED IN 1960.

Plant
Number of nests

Average height of plant

Average height of nest

Ulmus americana

4

7 ft. 6 in.

2 ft. 3 in.

Maclura pomifera

20

13 ft. 11 in.

1 ft. 11 in.

Crataegus mollis

1

11 ft.

3 ft. 1 in.

Gleditsia triacanthos

2

15 ft. 6 in.

1 ft. 9 in.

Acer negundo

4

8 ft. 9 in.

2 ft. 5 in.

Cornus drummondi

2

8 ft.

2 ft. 8 in.

Symphoricarpos orbiculatus

3

3 ft.

1 ft. 10 in.

7

36

11 ft. 4 in.

2 ft. 3 in.

Selection of the first nest-site may take as long as two days, possibly owing to incomplete development of the nesting tendency, but more likely to a general lack of familiarity with the territory.

Red-eyed Vireos require five to six days to choose the first nest-site (Lawrence, 1953:54). Later sites of the Bell Vireo are chosen in as little as three hours. Nest 1-c (1960) was abandoned at about 11:00 a.m. on May

14, 1960, when part of the thicket on the edge of which this nest was located was removed by brush-cutters clearing a power line right-of-way. By 2:00 p.m. this pair had begun construction of 1-d (1960) in an Osage orange 110 feet southwest of 1-c (1960).

This particular site is of further interest because it is the same one utilized for nest 1-a (1960). In all, four instances of utilization of a nest-site a second time were recorded. Two-a (1960) and 2-d (1960) were built in the same fork; 1-c (1960) and 1-h (1960) were in the same tree, but not the same fork. It should be mentioned that 1-a (1960) and 2-a (1960) were abortive attempts that did not progress beyond the suspension apparatus. Nice (1929:16) recorded a similar instance of the re-use of a nest tree, but different forks were used.

Re-use of an exact nest-site would ordinarily be impossible if the initial attempt were not abortive, because the presence of a completed nest would pose problems in construction with which the birds would probably be unable to cope. (A report by Morse in Bent, 1950:256 of a double nest indicates that this may not always be true. At the time of discovery one nest contained two eggs and the other nest contained young.) Since nests are used only once there would be no tendency to adopt the old nest. However, abortive nests, usually little more than a few strands of nesting material secured to the fork, might stimulate the birds to continue building. Re-use of a single nest-site in 15.8 per cent of 38 nests built in 1960 seems to be more than fortuitous circumstance. This re-use may have physiological benefits in conjunction with apportionment of energy for other nesting activities, because rapid location of a nest-site would mean that energy normally expended in searching and selecting could be rechanneled for actual construction.

In each of the instances of rebuilding, the new nest was begun on the same day that the previous nest was abandoned.

The renesting of pair 9 (1960) is worthy of note. These birds were established in the elm thicket on Clark land. Elm was by far the most abundant tree, with dogwood, Osage orange and honey locust also relatively common. There were only six boxelders in the territory and yet the four nests built by this pair were placed in them. This is the only instance of seeming preference.

Building

Nestbuilding by Bell Vireos can be best discussed in terms of the phases of construction described for the Red-eyed Vireo, Lawrence (1953:57), which are: (1) construction of the suspension

apparatus, (2) construction of the bag, (3) lining of the bag and smoothing and polishing of the exterior, and (4) adornment of the exterior. Red-eyes (Lawrence, 1953:59) may continue adornment far into the period of incubation. Both the male and female Bell Vireo have been observed to add spider egg sacs and other silk to the exterior of the nest as late as the sixth day of incubation.

Nice (1929:16) recorded only the female Bell Vireo building, but she did recall, from previous studies, having seen males aiding somewhat. Pitelka and Koestner (1942:102) wrongly concluded that the female Bell Vireo builds unaided, but Hensley (1950:243) observed that both sexes participated in nestbuilding, and Mumford (1952:229) reported two instances of building by both adults.

His description of the activities viewed in mid-May suggest that they were of the transitional period between the first and second phases. On the second occasion he recorded both adults building during the second phase. Since no details accompany this second observation I assume that it pertained to activity not necessarily typical of this phase of construction. Whereas both sexes of the Bell Vireo cooperate in building the nest, only the female Red-eyed Vireo builds according to Lawrence (1953:56). But Common

(1934:242) saw both Red-eyed Vireos building a nest.

The suspension apparatus is constructed by only the male on the first day. He punctuates each trip to the nest with song. The single song phrase is given from three to eight times when the male, carrying nesting material in his bill, arrives in the tree. Typically, he alights on several perches within the nest tree before flying to the nest. He often interrupts his work with several songs; when he has finished adding a load of material he sings from several perches within the nest tree before departing. The male periodically stops building to court the female.

In eight hours (494 minutes) of observing the first phase of construction at five different nests, I saw the female come to the nest 28 times; the male made 95 trips. The female came alone only once, and brought nesting material ten times, but did not build; on the other 18 occasions her visits were brief and she usually confined her activities to an inspection of the nest. Twenty of the visits by the female were made late in the first phase, marking a gradual transition to her assumption of building responsibility. (The delay by the female in beginning to build is puzzling; because all evidence indicates that she helps select the nest-site, I would expect her to help with the initial building. There seems to be no clear advantage in her delay in beginning to build.) The courtship and building activities of the male plus the presence of a partly completed nest seem to stimulate the female to commence building. Her visits become more frequent as construction of the suspension apparatus nears completion. At a time early in the second day the transition has taken place, and the female becomes the sole worker.

On May 7, 1960, male 2 (1960), at the time unmated, was observed as he came upon a nest of the previous year. The nest, after a year's weathering, suggested in appearance perhaps an early second-day nest. The bird flew to the nest and tugged and wove loose strands of grass for three minutes. Before leaving the site, the bird sang twice from different perches. This observation suggests that a partly constructed nest can elicit nestbuilding behavior, even in an unmated male.

The techniques of building by the male consist primarily of laying pieces of grass or bark across the fork, or along one of its branches, and then fastening them in place with pieces of animal silk. Once a "racket" has been formed, spider egg cases and plant down are emplaced among the fibers. The male employs weaving, twisting, and pecking motions of the head to emplace material.

As previously indicated, the female is the principal worker in the second and third phases of construction. The male infrequently visits the nest, but regularly visits the nest tree. The molding of the bag is accomplished by piling leaves, grasses and plant down onto the suspension apparatus. This material is also bound in with animal silk. As the amount of material accumulates, the female begins to trample it and gradually the bag takes shape. When trampling is first attempted, the nest often fails to support the female and she falls through the bottom of the nest. Such an occurrence was observed on May 23, 1960, on three consecutive trips by female 1

(1960), in constructing nest 1-e (1960). As the bag deepens, additional strands of grass are added to the wall and woven into place.

The male is extremely attentive during this and the following phase.

He follows the female as she gathers nest-material accompanying both this activity and her building with rapid song; he may give an average of seven song phrases per minute. The male brings to the nest a strand of grass, or some other material, about every twentieth trip. He frequently inspects the nest and the activities of the female from perches near the nest. Construction of the bag is ordinarily completed in the third day.

The third phase, the lining of the interior and the smoothing of the exterior, involves an additional one and one-half to two days.

Smoothing of the exterior refers to tightening of the grasses woven into the bag and addition of more animal silk. In lining the nest, the female stands on one of the branches of the fork and emplaces one end of a long, thin strand of some relatively stiff piece of grass or strip of bark. She then jumps into the bag and, while slowly turning around, pecks it into place, thus coiling the strand neatly around the interior of the bag.

As previously mentioned, the fourth phase overlaps the periods of lining, smoothing, egglaying, and incubation. The principal activity is the addition of white spider egg sacs to the exterior.

The trips are infrequent; but, occasionally, birds will interrupt an hour of incubation with three or four minutes of active adornment, during which several trips may be made. Both sexes participate in this phase.

Gathering of Nesting Material

Nesting materials were gathered anywhere within the territory.

Occasionally materials were collected from within the nest tree, but usually they were obtained 20 to 200 feet from the nest-site.

On several occasions I observed birds inspecting stems or branches where bark was frayed. Loose ends are grasped in the beak and torn free with an upward jerk of the head. Possibly the notch near the distal end of the upper mandible aids in grasping these strands.

Plant down is first extracted and then rolled into a ball by means of the beak while held with the feet before being transported to the nest.

Length and Hours of Nestbuilding

As indicated by Nolan (1960:230), accurate determination of the length of nestbuilding is difficult because of continued adornment and polishing after the nest is functionally complete. Most of the early nests for which I have records took from four and one-half to five days to construct. A four-to five-day period of building is reported by other observers (Nice, 1929:16; Pitelka and Koestner, 1942:99; Hensley, 1950:242; Nolan, 1960:230).

One instance of protracted building was recorded. Nest 6-d (1960) was begun on May 29, 1960, and not completed until nine days later on June 6, 1960. In contrast nest 1-g (1960) begun on May 31, 1960, was finished three days later on June 2, 1960. Nestbuilding occurs between the hours of 6:00 a.m. and 5:30 p.m. Heavy rain in the early morning may delay building.

Abortive Nestbuilding Efforts

Eight of 38 nests started in 1960 were never completed (Table 6).

Six of these abortive attempts were abandoned during, or shortly after, the completion of the suspension apparatus. Five of these nests were abandoned because the female did not begin building following the end of work by the male. The early abandonment of the other three nests 1-a (1960), 2-c (1960) and 6-e (1960) was attributable to the interruption of building by the male because of heavy rain and protracted territorial conflicts. The occurrence of these abortive nests at any time within the nesting efforts of a single pair indicates that such attempts are not examples of "false nestbuilding."

Renesting

Renesting after desertion or successful fledging occurs within two to thirty-six hours. Young were fledged from 1-a (1959) on June 19, 1959, and nest 1-b (1959) was discovered when late in the second phase of construction on June 22. If the nest was started on June 20, then renesting took place within 15 hours after fledging.

The Nest

Several authors have described various aspects of the nest of the Bell Vireo, notably Goss (1891:535); Simmons (*in* Bent, 1950:256), Nice (1929:13) and Nolan (1960:230-231). I can add but little to these descriptions.

The nest itself is a compact structure composed of strips of bark and strands of grasses that are interwoven and tightly bound with animal silk. The floor of the cup is first lined with a layer of small leaves and then the entire interior is lined with fine stems or strips of bark. Feathers are occasionally used to pad the bottom prior to lining, as are pieces of wool and milkweed down. Nest 2-e (1960) had been packed with small pieces of soil bearing moss prior to lining.

TABLE 6. ABORTIVE NESTING ATTEMPTS IN MAY AND JUNE OF 1960.

Nest	Length of time worked on	Cause of abandonment
1-a	1 day	Heavy rain
1-h	2 days	Female failed to build
2-a	1/2 day	Female failed to build
2-c	1 day	Protracted territorial dispute
4-a	1 day	Female failed to build
5-a	1 day	Female failed to build
6-c	1 day	Heavy rain
7-a	2 days	Female failed to build

Early nests tend to be bulkier, having thicker walls and bottoms than later efforts. However, nests in May were found to have 16

per cent thicker bottoms and 41 per cent thicker walls than nests in June (Table 7). Standard nest measurements do not show this to be so, for the exterior and interior diameters at the rim are governed by the angle between the two branches of the fork.

TABLE 7. DIMENSIONS OF NESTS IN MAY (1960) AND JUNE (1960).

Measurements
May (N 10)
June (N 8)
External depth
61.6 mm.
59.3 mm.
Depth of cup
45.5 mm.
46.3 mm.
Outside diameter
57.3/55.5 mm.
54.3/53.5 mm.
Inside diameter
43.4/42.2 mm.
45.5/43.9 mm.
Thickness of forward wall 1 inch below rim
13.8 mm.
7.6 mm.
Thickness of bottom
11.3 mm.
4.6 mm.

EGGLAYING AND INCUBATION

Egglaying

Egglaying begins the first or second day after completion of the nest. The female sits in the nest occasionally for periods of five to twenty-five minutes on the day the nest is completed. This is interrupted by periods of nest-adornment and foraging; such activities sometimes keep the female off the nest for several hours. Prior to the laying of the first egg, only the female is seen on the

nest, although the male is often seen sitting quietly within the nest tree a few feet from the female. The infrequency of the "congested" song and the alarm (*eh-eH-EH*) after the inception of "broodiness" indicates the waning of courtship behavior. As later in incubation only the "normal" song and the scold are heard.

Eggs are laid early in the morning prior to 5:30 a. m. according to Nolan (1960:232). The nest is usually left unoccupied for considerable periods after the first egg is laid, but, on the first day of laying, both sexes have been observed sitting for brief periods averaging ten minutes in length. Eggs are laid at one-day intervals until completion of the clutch. I found incubation to begin with the second egg.

Clutch-size

The average clutch-size of the Bell Vireo in Kansas, based on thirty-three records, is 3.39 eggs (Table 8). Seasonally, the largest average clutches are produced in the middle of the breeding season, that is, in June. Lack (1947:308-309) indicates that in European passerines the highest seasonal average clutch-sizes likewise occur in June. The largest average clutch-size in the Bell Vireo is presumably related to some aspect of the availability of food.

TABLE 8. AVERAGE NUMBERS OF EGGS PER NEST (NUMBER OF RECORDS IN PARENTHESES)[F].

Year

May

June

July

Mean annual clutch-size

1959

1960

3.0 (7)

3.3 (6)

3.2 (12)

3.83 (5)

3.0 (1)

4.0 (2)

3.06

3.72

1959-1960

3.17

3.52

3.5

3.39

[F] These data have been supplemented from the literature pertinent to Kansas.

Caution is necessary in determining mean clutch-size in the Bell Vireo. Eggs occasionally disappear from the nest prior to or during incubation, without subsequent addition of cowbird eggs. Unfamiliarity with the history of such a nest on the part of the observer would lead to an inaccurate determination of clutch-size.

Complete clutches are not replaced with the same regularity as are nests. I have recorded intervals of six to thirty days between successive clutches. Successful replacement of clutches is determined by a number of factors: nest-site, completion of a nest,

weather, predation, and parasitism by the cowbird. The difference between the number of renesting attempts and the successful replacement of clutches seems to indicate that different physiological processes are responsible for these two phenomena and that there is lack of synchrony between them. The development of the ovarian follicle requires a specific number of days that is not always coincident with the building of replacement nests. If, in the Bell Vireo, replacing a nest were solely a responsibility of the female, instead of involving the male to a considerable extent, it would seem likely that replacement of nests and the replacement of clutches would be more closely coordinated.

Incubation

Nice (1954:173) considers the incubation period to be the elapsed time between the laying of the last egg in a clutch and the hatching of that egg, when all eggs hatch. My data indicate that, normally, intensive incubation begins when the second egg is laid and lasts fourteen days in the Bell Vireo. Nice (1929:99) also considered the incubation period in this species to be fourteen days but believed it to commence when the third egg was laid. Pitelka and Koestner (1942:99) noted that the first and second eggs hatched fourteen days after laying of the second egg. However, they thought incubation began with the first egg. This would mean a fifteen-day period for this egg.

All the eggs that Nolan (1960:234) marked hatched in approximately fourteen days. Eight eggs artificially incubated by Graber (1955:103) required an average of 15.01 days to hatch. As Van Tyne and Berger (1959:293) indicate, periods of sitting on the nest, even all night, do not necessarily mean that incubation has begun, for it has been demonstrated in several species that birds may sit on an egg without actually applying heat. My own observations demonstrate that the first egg may be left unattended for several hours at a time on the day that it is laid.

The Roles of the Sexes in Incubation

Both the male and female sit on the eggs in the daytime. My study of histological sections of ventral epidermis indicates that the male does not possess a brood patch; the increased vascularization typical of the brood patch in females is not evident in males. But, the male loses most of the down feathers of the ventral apterium. Also, according to Bailey (1952:128), the male Warbling Vireo that sits on the eggs lacks a brood patch.

Bailey (1952:128) suggests that male passerines lacking brood patches that habitually sit on eggs do not heat the eggs. Thus it cannot be considered true incubation since no increase of temperature in the eggs is effected by such means. He further notes that it is at night when eggs are likely to experience a drop in temperature that embryonic development will be impaired. I have no data directly pertaining to which sex sits at night, but it is presumably the female, because she is always seen on the nest early in the morning and late in the evening.

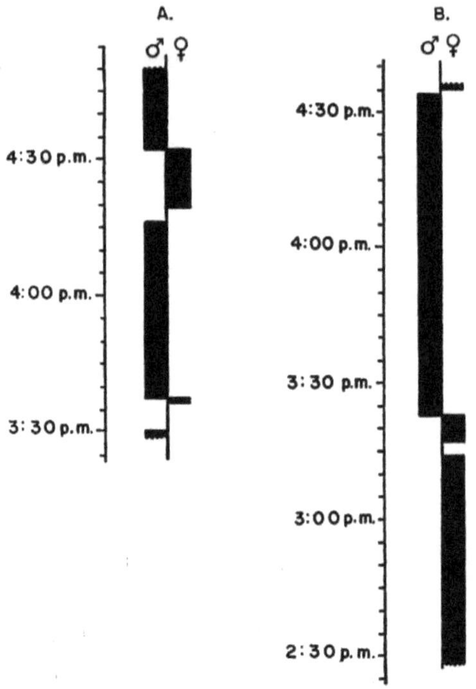

FIG. 4. Comparison of periods of incubation by both sexes in cold (54deg F.) rainy weather (A) and in warm (82deg F.) sunny weather (B).

If a highly-vascularized brood patch is essential for true incubation, then it is surprising that males take regular turns on the nest in cold, rainy weather. On May 20, 1960, male 3 (1960) sat on the eggs longer than did the female (fig. 4). The temperature during this hour and a half of incubation was 54deg F. One solution to this problem is supplied by Skutch (1957:74). He indicates that, "the type of the incubation is determined largely by innate factors, so that it persists through fairly wide fluctuations in weather, although it may break down in extreme conditions." Obviously then, in the example described above, the weather conditions do not qualify as "extreme." Sitting by the male is certainly functional to some extent for it relieves the female to forage; furthermore, the eggs are sheltered from inclement weather and protected from predators. Nolan (1960:232) suggests similar reasons for incubating by the male and adds the "conservation of heat supplied to the eggs by the female."

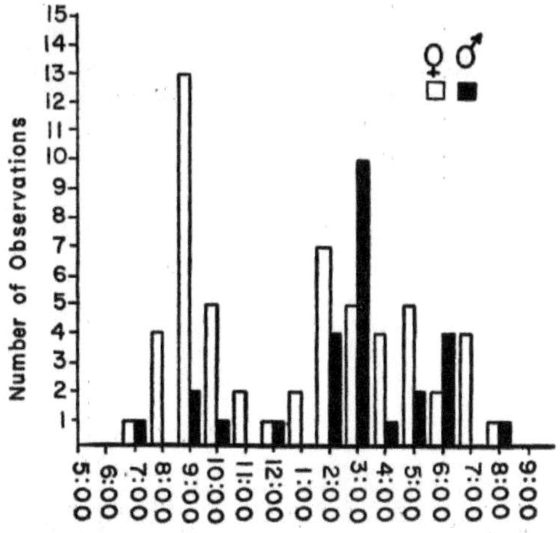

FIG. 5. Daily participation in incubation as indicated by the sex of the adult on the nest upon approach of the observer.

My data, based on incubation beginning with the second egg, indicate that the female incubates more often daily than the male (fig. 5). The male sits on the eggs only occasionally in the morning, but almost as often as the female in the afternoon. Nolan (1960:233) found that 95.5 per cent of the male's time on the nest and only 40 per cent of the female's time were attributable to the early hours of the day.

Although I lack data on the critical hours of 5:00 a.m. to 6:59 a.m., I have enough observations (20) from 7:00 a.m. to 9:00 a.m. to indicate that the males sit on the eggs infrequently (3 of 20 instances) in those hours. The discrepancy in the two sets of data, which may be merely an artifact of sampling techniques, does suggest two possible alternatives: (1) the male sits on the eggs in the morning and gives the female, who sits on the eggs throughout the night, an extended rest and an opportunity to forage; (2) the female continues to sit throughout the morning, especially during the early hours of daylight, a time of day when the temperature may still be low enough to impair development of the embryo.

Relief of Partners in Incubation

Relief of partners involves some ceremony. When the female is incubating, the male sings several times as he approaches the nest tree; the female responds with several *chees*, but otherwise remains immobile. The male sings several more times upon alighting in the nest tree whereupon the female *chees* again and flies directly from the nest. A few seconds later the male appears at the edge of the nest and, after inspecting the eggs, hops in and settles upon them.

When the male is sitting he is notably anxious prior to an exchange with the female, often arising and craning his neck as he surveys the surrounding vegetation, seemingly searching for his mate. The singing of the male and the calling of the female serve as signals, coordinating the exchange.

NESTLING PERIOD

Hatching Sequence

As indicated earlier, hatching normally occurs fourteen days after the second egg is laid. Hatching of the young was staggered at three nests under observation. In nest 2-b (1959) the first young hatched on June 8, 1959, the second on June 10. In 3-b (1959) one young hatched each day from the 12th through the 14th of June. In 5-a (1959) two young hatched on June 15, the third on June 16, and the fourth on June 17. Size of the young differed notably for about three days as a result of staggered hatching, but after that day the younger birds tended to catch up in size with their older brood-mates. The fourth young in nest 5-a (1959) grew steadily weaker and was missing from the nest on June 23, 1959. Staggered hatching is usually thought to be related to the availability of food that will insure survival of at least some of the nestlings when a shortage of food exists. It is doubtful that staggered hatching has adaptive significance in the Bell Vireo, since there seems to be no shortage of food for the young. In small passerines such as the Bell Vireo the principal problem is to insure fledging as quickly as possible because of the danger from predators.

Development of the Nestlings

Young are pinkish at hatching and devoid of visible natal down.

Du Bois (*in* Wetherbee, 1957:380), inspected day-old nestlings by means of a magnifying glass and was unable to detect any down.

Nolan (1960:236) also indicates that the young are naked at birth and that the "body color is between flesh and rufous except where folds of the straw yellow skin obscure the underlying colors."

The Hutton Vireo (*Vireo huttoni*) is essentially naked at birth, save for sparse hairlike down on the head and back (Wetherbee, 1953:380). The Red-eyed Vireo, according to Lawrence (1953:67) is naked at birth save for a sparse covering of greyish natal down, on the head, shoulders, and back.

In the Bell Vireo the pterylae darken slightly on the second day and the color becomes more intense daily until the quills of the dorsal tracts, the wings, and the tail break from their sheaths on the sixth day. In Red-eyed Vireos the pterylae darken by the end of the first day and the quills break through the skin on the fifth day, erupting from the sheaths by the seventh day (Lawrence, 1953:67).

From the first day the young are able to squeak. Poking a young bird was sufficient to elicit this sound, phonetically a nasal *peek.*

The only other vocalization noted throughout the nestling period was an abbreviated *chee.*

For the first three days tapping the nest or even movement of it caused by wind would elicit begging. By the fifth day at nest 2-a (1959) only vigorous agitation of the branch to which the nest was attached evoked any response. At this nest on June 16, 1959, one young begged while the other cowered. Cowering is correlated with opening of the eyes, as the young bird that begged had its eyes only partly open. Both young cowered on June 19, 1959.

Table 9 summarizes the maturation of the nestling Bell Vireos.

TABLE 9. MATURATION OF NESTLING BELL VIREOS. THE FIRST DAY THAT AN ACTIVITY WAS OBSERVED IS SHOWN.

Day of nestling life

1
2
3
4
5
6
7
8
9
10
11
Eyes open

x

Feathers erupt

x

Sound: Squeak

x

Sound: *Chee*

x

Begging

Cowering

x

Head scratching and Preening

x

Hopping to rim of nest

x

Fledging

x[G]

[G] This is the commonest fledging day.

Parental Behavior

No eggshells were found in nests on the days of hatching. Presumably they had been removed by the parents. Nolan (1960:234) indicates immediate disposition of the eggshell after hatching. Lawrence (1953:62) suggests that conspicuous removal of eggshells by the female Red-eyed Vireo informs the male that the young have hatched.

Both sexes brood and the exchange of partners resembles that described for the incubation period. Decrease in brooding in the daytime begins about the sixth day of nestling life. Nolan (1960:235) reports a sharp decrease in brooding when the oldest nestlings are seven days old. Brooding decreases notably on the sixth day of nestling life in the Red-eyed Vireo (Lawrence, 1953:62). Nice (1929:17), Hensley (1950:244), and Nolan (1960:235) report that the female Bell Vireo assumes a slightly greater role in brooding than the male.

Apparent sun-shading was noted at nest 3-b (1959) at 2:00 p.m. on June 17, 1959, on the fifth day of the nestling period. The nest contained three young. An adult flew to the nest; while standing on its rim the bird dipped its head into the nest six times, afterward appeared to be eating a fecal sac, than shifted position to the unattached portion of the rim, gaped three times, thereupon spread its wings, and sat motionless 35 minutes. In this attitude it formed an effective shield sheltering the young from direct sunlight penetrating the thin foliage of the honey locust in which the nest was situated. The temperature at this time was 95deg F., but the sky was partly cloudy. By 2:30 p.m. the sky had become overcast and the sun passed behind a cloud. Although sunlight no longer fell directly upon the nest, the bird remained in the shielding posture for another five minutes before flying from its perch. Sun-shading was not observed at either of the other nests containing young; dense overhead vegetation protected those nests.

Sun-shading has been noted in other species where the nest was poorly protected from the sun. Lawrence (1953:62) observed this behavior at two Red-eyed Vireo nests in conifers. The "sun-shield" posture of the Bell Vireo does not correspond to any of the sunning postures described by Hauser (1957).

Feeding of the Nestlings

Both sexes fed the young, and presumably began shortly after the first nestling hatched. My data indicate that the female does more feeding than the male (Table 10); in about eight hours of observation a total of 67 morsels were brought, 43 by the female and 24 by the male, for an average of once every 7.6 minutes. Nice (1929:17), however, observed a male to bring food 53 times as compared to 21 visits by the female. In five and one-half hours of watching, meals were brought once every 4.9 minutes. Du Bois (*in* Bent, 1950:257) recorded seven trips in an hour and forty minutes, or one every fourteen minutes.

At three nests containing young the adults were sometimes silent and sometimes vocal on their approach. The female often emitted a subdued *chee* which, coupled with the vibration of the nest caused by her arrival, elicited begging behavior from the young. None of the males was heard to utter such a call, but I have the impression that they often did call although I failed to hear the sounds. The males did, on occasion, sing several songs as they approached, even with food held in their beaks. Such singing elicited begging from the nestlings. Once the eyes of the young were open they often began begging when a silent adult was within two or three feet of the nest; begging behavior probably is elicited by tactile, auditory or visual stimuli in that order, or, as the nestling period proceeds, by any combination of these stimuli.

TABLE 10. FEEDING OF THE NESTLINGS.

Day of nestling period	Length of observation	Adult involved	
		Male	Female
1	30 min.	3	5
2	60 min.	1	4
3	60 min.	2	5
4	30 min.	1	4
7	60 min.	4	7
2	60 min.	3	3
6	60 min.		

3

6

7

30 min.

3

3

9

60 min.

4

6

Totals

510 min.

24

43

Not all trips made by parents resulted in successful feeding of young; some visits seemed to be purely for inspecting the young.

On other occasions the adults experienced difficulty in transferring food to the young, and, thus thwarted, would themselves eat the food. Nice (1929:17) estimated that from five to twelve of a total of seventy-five meals were eaten by adults.

Nest Sanitation

Both parents regularly removed fecal sacs from the nest, eating them for the first five days and thereafter carrying them off and presumably dropping them. It is doubtful that fecal sacs were actively removed in the last two days of nestling life as the bottoms of nests from which young flew away were invariably covered with excrement.

On several occasions a parent brought food to the nest and then remained perched on the rim alternately peering into the nest and then preening. Once bill swiping was observed and another time an adult male sang once. The adult remained at the nest from twenty seconds to a full minute.

Fledging

Eight young were fledged from the four nests in 1959. The

nestling period lasted from nine to twelve days. Human interference may have been largely responsible for the fledging of the young at nine days. Pitelka and Koestner (1942:100) found nestling life to last eleven days. Nolan (1960:235) reports nestling periods varying from 10.5 to 12 days. The young Red-eyed Vireo is ready to leave the nest at ten days but often remains an additional day before departing (Lawrence, 1953:68).

The oldest nestling at nest 2-a (1959) hopped out on June 17, 1959, when I disturbed the parents. On this date the juvenal plumage was only partly developed and the young bird was incapable of flight. By the tenth day of nestling life the young in all the nests were observed to hop to the rim, flutter their wings, hop back into the nest and also to preen and scratch their heads. The young at fledging are usually completely feathered, but have notably short tails and relatively short, stubby wings. According to Ridgeway (1904:205) the juvenal plumage is much like that of the adult.

Nest Parasites·

Pitelka and Koestner (1942:103) found that incubating adults and later the young suffered infestation of the northern fowl mite,

Ornithonyseus sylviarum. Nolan (1960:241) reports a heavy infestation of this mite at four nests. Unidentified mites were noted at four nests in my study area in 1959. Incubating adults were observed to peck at their breasts and scapulars from the eleventh through the fourteenth day of incubation. Serious infestations were not noted at the nests until the ninth day of nestling life. At this time the young were observed to scratch their heads and peck at their breasts, scapulars, and the base of their tails. On the day of fledging the nests were a seething mass of crawling mites; the mites also extended well up the branches to which the nests were attached.

Nest 1-a (1959), which was discovered on June 18, 1959, presumably on the day after fledging, was densely covered with mites. Some mites were still crawling on this nest on June 20, 1959.

FLEDGLING LIFE

On June 20, 1959 I located one young 80 feet northeast of nest 2-a (1959), about five hours after it had left the nest. One parent was observed to feed it once. No young were seen thereafter from this or any other nest. Extreme agitation on the part of one or both parents on several occasions shortly thereafter, however, suggested the proximity of the young. Search in the immediate vicinity on each of these occasions proved fruitless. Three days after fledging their young, pair 2 (1959) was primarily occupied with courtship activities.

Pair 1 (1959) was involved in courtship and nestbuilding one and one-half days after the apparent fledging of their young. Nolan (1960:238) indicates that the young remain within the territory and perhaps are fed by the parents up until an age of about 40 days.

Sutton (1949:25) and Lawrence (1953:68) present contradictory reports on fledgling-parent relationships in the Red-eyed Vireo. Sutton concluded that the young quickly took leave of their parents whereas Lawrence reported a young bird being fed 35 days after fledging.

Second Broods

The curve based on 66 nesting records of the Bell Vireo representing the breeding activity in northeastern Kansas demonstrates

a tendency toward double-broodedness (fig. 6). The peak of the breeding season is from May 20 to June 20. The large number (20) of replacement nests built in late May of 1960 tends to distort the curve of the breeding data; a second peak about 35 days after the first is evident.

I am of the opinion that the vast majority of vireos are single-brooded solely by virtue of the limited success of early nesting

efforts, and that in "good" years most pairs would be double-brooded.

Each of the four pairs that successfully raised one brood

in my study area in 1959 renested within a day or two after the fledging of the young. I do not know the fate of these nests. Nolan (1960:237) reports at least one instance of a second brood in the course of his study. Nolan (*op. cit.*) notes that the literature, in general, indicates that vireos are double-brooded, but that his evidence, mentioned previously, is the only evidence based on banded birds.

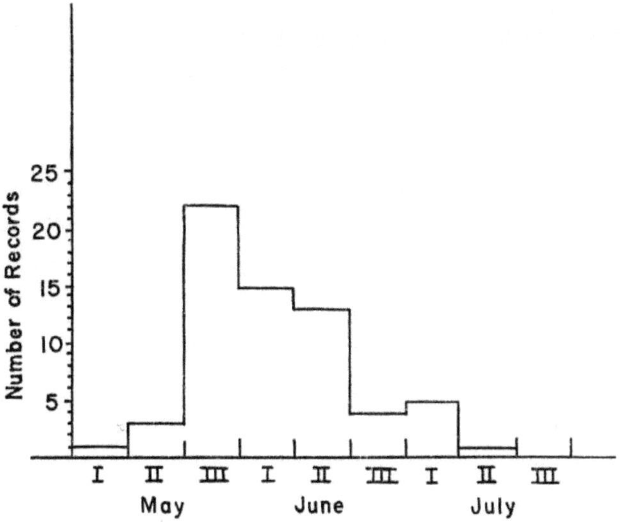

FIG. 6. Breeding season in northeastern Kansas based on the number of completed clutches in each 10-day period from May through July.

REPRODUCTIVE SUCCESS

Only four nests were successful; all of these were observed in 1959. The principal external factors responsible for nesting failure were severe weather, predation, parasitism by Brown-headed Cowbirds (*Molothrus ater*) and human interference (Table 11).

In late winter and early spring of 1960 heavy snow, continuously at a depth of at least 10 inches, covered most of the Mid-west from February 20 through March 20. Consequently, the growing season was some two weeks behind that of 1959. Of all the species in the study area, the Bell Vireo is the most dependent on dense foliage for cover and concealment for its nests. Consequently the tardiness of the season seemingly negatively influenced reproductive success of this more than any other species of bird in the study area.

Behavior

Several aspects of the behavior of the Bell Vireo tend to contribute to nesting failure. They include:

1. Nest-site. Nests are occasionally suspended from exposed branches. Occurrences of this sort suggest that the dimensions of the fork are more important in the choice of a site than availability of cover.

2. Song. The loud, continuous song of the male during nestbuilding alerts cowbirds and predators to the presence of a nest.

The incongruous habits of the male of singing in the nest tree and while sitting on the nest may facilitate location by some enemies, particularly cowbirds.

TABLE 11. EGG MORTALITY IN BELL VIREOS.

Mortality agents

N[H]

Eggs (N-29)

1959 Per cent

N

1960 Per cent

Predation

4

13.8

5

10

Weather

2

6.9

8

16

Cowbird

14

48.3

37

74

Totals

20

69[I]

50

100

[H] Number of eggs out of the total number laid lost to mortality agents.

[I] In 1959 nine eggs were successful (ultimately gave rise to fledglings).

I am not fully convinced that song from the nest is simply a "foolish" habit, since snakes, the principal predators with which this species has to contend, are deaf. My own field observations and the circumstances of the innumerable instances recorded in the literature of male vireos singing from the nest suggest that this is a function of the proximity of the observer. As mentioned elsewhere, vocal threat is the initial as well as the primary means by which territory is maintained. Song from the nest evoked by an enemy also serves to alert the female to danger.

3. Flushing. The Bell Vireo normally relies upon cryptic behavior to avoid detection at the nest. Most sitting birds, especially the females, either flush silently when an enemy is about forty feet from the nest or remain sitting upon the nest tenaciously, refusing to flush even when touched or picked up. Some birds flushed at intermediate distances of from three to fifteen feet. In so doing they revealed the location of their nests. Since none of these "intermediate flushers" enjoyed nesting success there is possibly some correlation between these two factors.

Predation

Several complete clutches being incubated disappeared from nests that were unharmed. Absence of eggshells in the vicinity suggests predation by snakes.

On May 25, 1960, I found a *Peromyscus* climbing toward nest 1-a (1960). The mouse moved to within two inches of the nest whereupon I removed the mouse. Such small rodents constitute another potential source of predation.

Cowbird Parasitism

In this study the failure of 12 of 35 nests can be directly attributed to cowbird interference. It is well established that the incidence of cowbird parasitism of Bell Vireo nests is high (Friedmann, 1929:237; Bent, 1950:260-261). Nolan (1960:240) found only one nest of eight studied to be parasitized by cowbirds. He indicates that this is surprising in view of the heavy molestation of the Prairie Warbler (*Dendroica discolor*) in the same region. A possible explanation of this phenomenon seems to lie in the much greater abundance of the Prairie Warbler in comparison to that of the Bell Vireo. In my study area the incidence of cowbird parasitism on Bell Vireos in 1959 and 1960 greatly exceeded that of all other nesting species that were parasitized (Table 12).

As indicated previously, the female Bell Vireo leaves the nest unoccupied several hours at a time in the transition period between completion of the nest and the start of egglaying. Such behavior early in the morning certainly would facilitate deposition of cowbird eggs. Early in the nesting period the mere presence of a

cowbird egg in the nest prior to the laying of the host's first egg leads to abandonment of the nest. This seems to be correlated with the relative strength of the nesting tendency; anyhow cowbird eggs laid in later nests prior to the appearance of the host's own eggs did not cause the nesting birds to desert. The Bell Vireo does abandon the nest when all but one of its own eggs have been removed by the cowbird. Mumford (1952:232) records the removal of a cowbird egg by the host birds and I recorded a similar instance

involving nest 2-b (1960). On May 14, 1960, I found one punctured cowbird egg on the ground about 10 feet west of this nest. Occasionally a cowbird egg is buried beneath the lining of a nest. Mumford (1952:23) observed this in mid-May in 1951 and I observed pair 8 (1960) actively covering with building material a cowbird egg on July 5, 1960. Covering a cowbird egg constitutes effective removal. Since the egg cannot be turned, an adhesion develops.

TABLE 12. INCIDENCE OF COWBIRD PARASITISM OF THE BELL VIREO COMPARED WITH OTHER PASSERINES IN THE STUDY AREA IN 1959 AND 1960.

	Bell Vireo	Other passerines
Total nests examined containing at least one host egg	35	43
Total nests parasitized	24	14
Total number of cowbird eggs	33	23
Per cent of nests parasitized	68.6	32.6
Total number of cowbird eggs per nest	.94	.54

The percentage of cowbird eggs hatched in relation to the number laid is relatively low. For instance, Mumford (1952:231) has only one record of a young cowbird successfully raised by a Bell Vireo. The data available in Bent (1950:260-261) also indicate that the percentage of cowbird eggs hatched is small. The Bell Vireo is less tolerant of cowbird parasitism than are many of the species so victimized, but is not so intolerant as the Robin, Catbird, and the Yellow-breasted Chat (Friedmann, 1929:193).

SUMMARY

1. The behavior of a small population of Bell Vireos was studied in the spring and summer of 1959 and again in 1960 in Douglas County, Kansas, and results are compared with previous studies elsewhere.

2. The Bell Vireo sings more often daily and throughout the nesting season than do the majority of its avian nesting associates.

Six types of vocalizations are readily distinguishable in the field: primary song, courtship song, distress call, alarm note, specialized male call note or *zip*, and the generalized call note or *chee*.

3. Territories are established in early May and occupied throughout the breeding season and post-breeding season. The average size of the territories in 1960 was 1.25 acres. Shifting of territorial boundaries occasionally occurs after nesting attempts.

4. Territory is maintained primarily by song, but at least five aggressive displays are manifest in the early phases of territorial establishment. These include: (a) vocal threat, (b) head-forward threat, (c) wing-flicking and submaximal tail-fanning, (d) ruffling and maximum tail-fanning, and (e) supplanting attack.

5. The precise mechanism of pair-formation in the Bell Vireo is not known. Early courtship activities are characteristically violent affairs. Absence of sexual dimorphism suggests that behavioral criteria are used by the birds in sex-recognition; the male is dominant and the female is subordinate.

6. The principal displays associated with courtship include: greeting ceremonies, "pouncing," "leap-flutter," pre-and post-copulatory displays, and the posture, copulation. The marked similarity between elements of courtship display and aggressive display suggests common origin or the derivation of one from the other.

7. The nest-site probably is selected by the female. Nests are suspended from lateral or terminal forks about 2 feet 3 inches high in small trees and shrubs averaging 11 feet 2 inches in height.

8. Nestbuilding is intimately associated with courtship and is a responsibility of both sexes. The male builds the suspension apparatus and the female constructs and lines the bag. Both sexes participate in adorning the exterior. Construction lasts from four and one-half to five days.

9. The nest is compact, pendant, and composed of strips of bark and strands of grasses that are interwoven and tightly bound with animal silk. Nests built in May are bulkier than those constructed later in the season.

10. Egglaying begins on the first or second day after the nest is completed. The eggs are deposited early in the morning. The average clutch-size of the Bell Vireo in Kansas is 3.39 eggs.

11. Both sexes sit on the eggs, but only the female truly incubates because the male lacks a brood patch. Incubation lasts fourteen days.

12. The Bell Vireo is double-brooded in "good" years.

13. Nesting failure resulted from severe weather, predation, parasitism by cowbirds, and human interference. Behavior that contributes to nesting failure is selection of an unfavorable nest-site, singing on and near the nest, and the tendency to flush from the nest in view of potential enemies.

LITERATURE CITED

AMERICAN ORNITHOLOGISTS' UNION

1957. Check-list of North American birds. Fifth ed. Baltimore, The Lord Baltimore Press, The American Ornithologists' Union, iv + 691 pp.

ANDREW, R. J.

1956. Intention movements of flight in certain passerines, and their use in systematics. Behaviour, 10:79-204.

BAILEY, R. E.

1952. The incubation patch of passerine birds. Condor, 54:121-136.

BENNETT, W. W.

1917. Bell's Vireo studies (Vireo bellii Aud.). Proc. Iowa Acad. Sci., 24:285-293.

BENT, A. C.

1950. Life histories of North American wagtails, shrikes, vireos and their allies. Smithsonian Inst., U. S. Nat. Mus. Bull., 197:vii + 411 pp., 48 pls.

BUNKER, C. D.

1910. Habits of the Black-capt Vireo (Vireo atricapillus). Condor, 12:70-73.

CHAPIN, E. A.

1925. Food habits of the vireos; a family of insectivorous birds. U. S. Dept. Agric. Bull., 1355:1-44.

COMMON, M. A.

1934. Notes on a Red-eyed Vireo's nest. Auk, 51:241-242.

COOKE, W. W.

1909. The migration of vireos. Bird Lore, 11:78-82, 118-120, 165-168.

FITCH, H. S.

1958. Home ranges, territories and seasonal movements of vertebrates of the Natural History Reservation. Univ. of Kansas Publ. Mus. of Nat. Hist., 11:3:63-326.

FRIEDMANN, H.

1929. The Cowbirds. Charles C. Thomas, Springfield, Illinois, xviii + 421 pp.

GOSS, N. S.

1891. History of the birds of Kansas. Topeka, Geo. W. Crane & Co. Printers and Binders. 692 pp.

GRABER, R. R.

1955. Artificial incubation of some non-galliform eggs. Wilson Bull., 67:100-109.

HAMILTON, T. H.

1958. Adaptive variation in the genus Vireo. Wilson Bull., 70:307-346.

HAUSER, D. C.

1957. Some observations on sun-bathing in birds. Wilson Bull., 69:78-90.

1959. Notes on pairing and nestbuilding of mismatched vireos. Wilson Bull., 71:383-384.

HENSLEY, MAX

1950. Notes on the breeding behavior of the Bell's Vireo. Auk, 67:243-244.

HINDE, R. A.

1952. The behaviour of the Great Tit (Parus major) and some other related species. Leiden: E. J. Brill, x + 201 pp.

1956. The biological significance of the territories of birds, Ibis, 98:340-369.

HINDE, R. A., and TINBERGEN, N.

1958. The comparative study of species-specific behavior. In Behavior and Evolution (Yale University Press, New Haven), pp. 251-268.

KLUIJVER, H. N.

1951. The population ecology of the great tit, Parus m. major L. Ardea, 39:1-135.

LACK, D.

1947. The significance of clutch-size. Ibis, 89:302-352.

LAWRENCE, L. DE K.

1953. Nesting life and behavior of the Red-eyed Vireo. Can. Field-Nat., 67:46-77.

LEWIS, H. F.

1921. A nesting of the Philadelphia Vireo. Auk, 38:26-44, 185-202.

LINSDALE, J. M.

1928. Birds of a limited area in eastern Kansas. The Univ. of Kansas, Sci. Bull., 18:11:517-626.

LLOYD, W.

1887. Birds of Tom Green and Concho Counties, Texas. Auk, 4:181-193, 289-299.

MORRIS, D.

1956. The feather postures of birds and the problem of the origin of social signs. Behaviour, 9:75-113.

MOYNIHAN, M.

1955. Types of hostile display. Auk, 72:247-259.

MUMFORD, R. E.

1952. Bell's Vireo in Indiana. Wilson Bull., 64:224-233.

NICE, M. M.

1929. The fortunes of a pair of Bell Vireos. Condor, 31:13-20.

1943. Studies in the life history of the song sparrow. II. The behavior of the song sparrow and other passerines. Trans. Linn. Soc. N.Y., 6:328 pp.

1954. Problems of incubation periods in North American birds. Condor, 56:173-197.

NOLAN, V.

1960. Breeding behavior of the Bell Vireo in southern Indiana. Condor, 62:225-244.

PITELKA, F. A.

1959. Numbers, breeding schedule, and territoriality in Pectoral Sandpipers of northern Alaska. Condor, 61:223-264.

PITELKA, F. A., and KOESTNER, E. J.

1942. Breeding behavior of Bell's Vireo in Illinois. Wilson Bull., 54:97-106.

RIDGWAY, R.

1889. The ornithology of Illinois. Illinois State Nat. Hist. Survey, 1: viii + 520 pp.

1904. The birds of North and Middle America. U.S. Nat. Mus. Bull., 50, pt. 3; xx + 801 pp.

SKUTCH, A. F.

1957. The incubation patterns of birds. Ibis, 99:69-93.

SOUTHERN, W. E.

1958. Nesting of the Red-eyed Vireo in the Douglas Lake Region, Michigan. The Jack-Pine Warbler, 36:105-130, 185-207.

SUTTON, G. M.

1949. Studies of the nesting birds of the Edwin S. George Reserve. Part 1. The Vireos. Misc. Pub. Univ. Michigan Mus. Zool., 74:5-36.

TOWNSEND, C. W.

1920. Supplement to the birds of Essex County, Massachusetts. Mem. Nuttall Orn. Club, No. 5:196 pp.

TYLER, W. M.

1912. A vireo courtship. Bird Lore, 14:229-230.

VAN TYNE, J., and BERGER, A. J.

1959. Fundamentals of ornithology. John Wiley & Sons, Inc., New York. xi + 624 pp.

WETHERBEE, D. K.

1957. Natal plumages and downy pteryloses of passerine birds of North America. Bull. Amer. Mus. Nat. Hist., 113:5:339-436.

Transmitted November 8, 1961.

29-1506

UNIVERSITY OF KANSAS PUBLICATIONS
MUSEUM OF NATURAL HISTORY

Institutional libraries interested in publications exchange may obtain this series by addressing the Exchange Librarian, University of Kansas Library, Lawrence, Kansas. Copies for individuals, persons working in a particular field of study, may be obtained by addressing instead the Museum of Natural History, University of Kansas, Lawrence, Kansas. There is no provision for sale of this series by the University Library, which meets institutional requests, or by the Museum of Natural History, which meets the requests of individuals. However, when individuals request copies from the Museum, 25 cents should be included, for each separate number that is 100 pages or more in length, for the purpose of defraying the costs of wrapping and mailing.

* An asterisk designates those numbers of which the Museum's supply (not the Library's supply) is exhausted. Numbers published to date, in this series, are as follows:

Vol. 1. Nos. 1-26 and index. Pp. 1-638, 1946-1950.

*Vol. 2. (Complete) Mammals of Washington. By Walter W. Dalquest. Pp. 1-444, 140 figures in text. April 9, 1948.

Vol. 3. *1. The avifauna of Micronesia, its origin, evolution, and distribution. By Rollin H. Baker. Pp. 1-359, 16 figures in text. June 12, 1951.

*2. A quantitative study of the nocturnal migration of birds. By George H. Lowery, Jr. Pp. 361-472, 47 figures in text. June 29, 1951.

3. Phylogeny of the waxwings and allied birds. By M. Dale Arvey. Pp. 473-530, 49 figures in text, 13 tables. October 10, 1951.

4. Birds from the state of Veracruz, Mexico. By George H. Lowery, Jr., and Walter W. Dalquest. Pp. 531-649, 7 figures in text, 2 tables. October 10, 1951.

Index. Pp. 651-681.

*Vol 4. (Complete) American weasels. By E. Raymond Hall. Pp. 1-466, 41 plates, 31 figures in text. December 27, 1951.

Vol. 5. Nos. 1-37 and index. Pp. 1-676, 1951-1953.

*Vol 6. (Complete) Mammals of Utah, *taxonomy and distribution*. By Stephen D. Durrant. Pp. 1-549, 91 figures in text, 30 tables. August 10, 1952.

Vol. 7. *1. Mammals of Kansas. By E. Lendell Cockrum. Pp. 1-303, 73 figures in text, 37 tables. August 25, 1952.

2. Ecology of the opossum on a natural area in northeastern Kansas. By Henry S. Fitch and Lewis L. Sandidge. Pp. 305-338, 5 figures in text. August 24, 1953.

3. The silky pocket mice (Perognathus flavus) of Mexico. By Rollin H. Baker. Pp. 339-347, 1 figure in text. February 15, 1954.

4. North American jumping mice (Genus Zapus). By Phillip H. Krutzsch. Pp. 349-472, 47 figures in text, 4 tables. April 21, 1954.

5. Mammals from Southeastern Alaska. By Rollin H. Baker and James S. Findley. Pp. 473-477. April 21, 1954.

6. Distribution of Some Nebraskan Mammals. By J. Knox Jones, Jr. Pp. 479-487. April 21, 1954.

7. Subspeciation in the montane meadow mouse, Microtus montanus, in Wyoming and Colorado. By Sydney Anderson. Pp. 489-506, 2 figures in text. July 23, 1954.

8. A new subspecies of bat (Myotis velifer) from southeastern California and Arizona. By Terry A. Vaughan. Pp. 507-512. July 23, 1954.

9. Mammals of the San Gabriel mountains of California. By Terry A. Vaughan. Pp. 513-582, 1 figure in text, 12 tables. November 15, 1954.

10. A new bat (Genus Pipistrellus) from northeastern Mexico. By Rollin H. Baker. Pp. 583-586. November 15, 1954.

11. A new subspecies of pocket mouse from Kansas. By E. Raymond Hall. Pp. 587-590. November 15, 1954.

12. Geographic variation in the pocket gopher, Cratogeomys castanops, in Coahuila, Mexico. By Robert J. Russell and Rollin H. Baker. Pp. 591-

608. March 15, 1955.

13. A new cottontail (Sylvilagus floridanus) from northeastern Mexico. By Rollin H. Baker. Pp. 609-612. April 8, 1955.

14. Taxonomy and distribution of some American shrews. By James S. Findley. Pp. 613-618. June 10, 1955.

15. The pigmy woodrat, Neotoma goldmani, its distribution and systematic position. By Dennis G. Rainey and Rollin H. Baker. Pp. 619-624. 2 figures in text. June 10, 1955.

Index. Pp. 625-651.

Vol. 8. Nos. 1-10 and index. Pp. 1-675, 1954-1956.

Vol. 9. 1. Speciation of the wandering shrew. By James S. Findley. Pp. 1-68, 18 figures in text. December 10, 1955.

2. Additional records and extension of ranges of mammals from Utah. By Stephen D. Durrant, M. Raymond Lee, and Richard M. Hansen. Pp. 69-80. December 10, 1955.

3. A new long-eared myotis (Myotis evotis) from northeastern Mexico. By Rollin H. Baker and Howard J. Stains. Pp. 81-84. December 10, 1955.

4. Subspeciation in the meadow mouse, Microtus pennsylvanicus, in Wyoming. By Sydney Anderson. Pp. 85-104, 2 figures in text. May 10, 1956.

5. The condylarth genus Ellipsodon. By Robert W. Wilson. Pp. 105-116, 6 figures in text. May 19, 1956.

6. Additional remains of the multituberculate genus Eucosmodon. By Robert W. Wilson. Pp. 117-123, 10 figures in text. May 19, 1956.

7. Mammals of Coahuila, Mexico. By Rollin H. Baker. Pp. 125-335, 75 figures in text. June 15, 1956.

8. Comments on the taxonomic status of Apodemus peninsulae, with description of a new subspecies from North China. By J. Knox Jones, Jr. Pp. 337-346, 1 figure in text, 1 table. August 15, 1956.

9. Extensions of known ranges of Mexican bats. By Sydney Anderson. Pp. 347-351. August 15, 1956.

10. A new bat (Genus Leptonycteris) from Coahuila. By Howard J. Stains. Pp. 353-356. January 21, 1957.

11. A new species of pocket gopher (Genus Pappogeomys) from Jalisco, Mexico. By Robert J. Russell. Pp. 357-361. January 21, 1957.

12. Geographic variation in the pocket gopher, Thomomys bottae, in Colorado. By Phillip M. Youngman. Pp. 363-387, 7 figures in text. February 21, 1958.

13. New bog lemming (genus Synaptomys) from Nebraska. By J. Knox Jones, Jr. Pp. 385-388. May 12, 1958.

14. Pleistocene bats from San Josecito Cave, Nuevo Leon, Mexico. By J. Knox Jones, Jr. Pp. 389-396. December 19, 1958.

15. New subspecies of the rodent Baiomys from Central America. By Robert L. Packard. Pp. 397-404. December 19, 1958.

16. Mammals of the Grand Mesa, Colorado. By Sydney Anderson. Pp. 405-414, 1 figure in text, May 20, 1959.

17. Distribution, variation, and relationships of the montane vole, Microtus montanus. By Sydney Anderson. Pp. 415-511, 12 figures in text, 2 tables. August 1, 1959.

18. Conspecificity of two pocket mice, Perognathus goldmani and P. artus. By E. Raymond Hall and Marilyn Bailey Ogilvie. Pp. 513-518, 1 map. January 14, 1960.

19. Records of harvest mice, Reithrodontomys, from Central America, with description of a new subspecies from Nicaragua. By Sydney Anderson and J. Knox Jones, Jr. Pp. 519-529. January 14, 1960.

20. Small carnivores from San Josecito Cave (Pleistocene), Nuevo Leon, Mexico. By E. Raymond Hall. Pp. 531-538, 1 figure in text. January 14,

1960.

21. Pleistocene pocket gophers from San Josecito Cave, Nuevo Leon, Mexico. By Robert J. Russell. Pp. 539-548, 1 figure in text. January 14, 1960.

22. Review of the insectivores of Korea. By J. Knox Jones, Jr., and David H. Johnson. Pp. 549-578. February 23, 1960.

23. Speciation and evolution of the pygmy mice, genus Baiomys. By Robert L. Packard. Pp. 579-670, 4 plates, 12 figures in text. June 16, 1960.

Index. Pp. 671-690.

Vol. 10. 1. Studies of birds killed in nocturnal migration. By Harrison B. Tordoff and Robert M. Mengel. Pp. 1-44, 6 figures in text, 2 tables. September 12, 1956.

2. Comparative breeding behavior of Ammospiza caudacuta and A. maritima. By Glen E. Woolfenden. Pp. 45-75, 6 plates, 1 figure. December 20, 1956.

3. The forest habitat of the University of Kansas Natural History Reservation. By Henry S. Fitch and Ronald R. McGregor. Pp. 77-127, 2 plates, 7 figures in text, 4 tables. December 31, 1956.

4. Aspects of reproduction and development in the prairie vole (Microtus ochrogaster). By Henry S. Fitch. Pp. 129-161, 8 figures in text, 4 tables. December 19, 1957.

5. Birds found on the Arctic slope of northern Alaska. By James W. Bee. Pp. 163-211, plates 9-10, 1 figure in text. March 12, 1958.

6. The wood rats of Colorado: distribution and ecology. By Robert B. Finley, Jr. Pp. 213-552, 34 plates, 8 figures in text, 35 tables. November 7, 1958.

7. Home ranges and movements of the eastern cottontail in Kansas. By Donald W. Janes. Pp. 553-572, 4 plates, 3 figures in text. May 4, 1959.

8. Natural history of the salamander, Aneides hardyi. By Richard F.

Johnston and Gerhard A. Schad. Pp. 573-585. October 8, 1959.

9. A new subspecies of lizard, Cnemidophorus sacki, from Michoacan, Mexico. By William E. Duellman, Pp. 587-598, 2 figures in text. May 2, 1960.

10. A taxonomic study of the middle American snake, Pituophis deppei. By William E. Duellman. Pp. 599-610, 1 plate, 1 figure in text. May 2, 1960.

Index. Pp. 611-626.

Vol. 11. 1. The systematic status of the colubrid snake, Leptodeira discolor Gunther. By William E. Duellman. Pp. 1-9, 4 figures. July 14, 1958.

2. Natural history of the six-lined racerunner, Cnemidophorus sexlineatus. By Henry S. Fitch. Pp. 11-62, 9 figures, 9 tables. September 19, 1958.

3. Home ranges, territories, and seasonal movements of vertebrates of the Natural History Reservation. By Henry S. Fitch. Pp. 63-326, 6 plates, 24 figures in text, 3 tables. December 12, 1958.

4. A new snake of the genus Geophis from Chihuahua, Mexico. By John M. Legler. Pp. 327-334, 2 figures in text. January 28, 1959.

5. A new tortoise, genus Gopherus, from north-central Mexico. By John M. Legler. Pp. 335-343. April 24, 1959.

6. Fishes of Chautauqua, Cowley and Elk counties, Kansas. By Artie L. Metcalf. Pp. 345-400, 2 plates, 2 figures in text, 10 tables. May 6, 1959.

7. Fishes of the Big Blue river basin, Kansas. By W. L. Minckley. Pp. 401-442. 2 plates, 4 figures in text, 5 tables. May 8, 1959.

8. Birds from Coahuila, Mexico. By Emil K. Urban. Pp. 443-516. August 1, 1959.

9. Description of a new softshell turtle from the southeastern United States. By Robert G. Webb. Pp. 517-525, 2 plates, 1 figure in text. August

14, 1959.

10. Natural history of the ornate box turtle, Terrapene ornata ornata Agassiz. By John M. Legler. Pp. 527-669, 16 pls., 29 figures in text. March 7, 1960.

Index Pp. 671-703.

Vol. 12. 1. Functional morphology of three bats: Eumops, Myotis, Macrotus. By Terry A. Vaughan. Pp. 1-153, 4 plates, 24 figures in text. July 8, 1959.

2. The ancestry of modern Amphibia: a review of the evidence. By Theodore H. Eaton, Jr. Pp. 155-180, 10 figures in text. July 10, 1959.

3. The baculum in microtine rodents. By Sydney Anderson. Pp. 181-216, 49 figures in text. February 19, 1960.

4. A new order of fishlike Amphibia from the Pennsylvanian of Kansas. By Theodore H. Eaton, Jr., and Peggy Lou Stewart. Pp. 217-240, 12 figures in text. May 2, 1960.

5. Natural history of the bell vireo. By Jon C. Barlow. Pp. 241-296, 6 figures in text. March 7, 1962.

More numbers will appear in volume 12.

Vol. 13. 1. Five natural hybrid combinations in minnows (Cyprinidae). By Frank B. Cross and W. L. Minckley. Pp. 1-18. June 1, 1960.

2. A distributional study of the amphibians of the Isthmus of Tehuantepec, Mexico. By William E. Duellman. Pp. 19-72, pls. 1-8, 3 figures in text. August 16, 1960.

3. A new subspecies of the slider turtle (Pseudemys scripta) from Coahuila, Mexico. By John M. Legler. Pp. 73-84, pls. 9-12, 3 figures in text. August 16, 1960.

4. Autecology of the copperhead. By Henry S. Fitch. Pp. 85-288, pls. 13-20, 26 figures in text. November 30, 1960.

5. Occurrence of the garter snake, Thamnophis sirtalis, in the Great

Plains and Rocky Mountains. By Henry S. Fitch and T. Paul Maslin. Pp. 289-308, 4 figures in text. February 10, 1961.

6. Fishes of the Wakarusa river in Kansas. By James E. Deacon and Artie L. Metcalf. Pp. 309-322, 1 figure in text. February 10, 1961.

7. Geographic variation in the North American cyprinid fish. Hybopsis gracilis. By Leonard J. Olund and Frank B. Cross. Pp. 323-348, pls. 21-24, 2 figures in text. February 10, 1961.

8. Descriptions of two species of frogs, genus Ptychohyla; studies of American hylid frogs, V. By William E. Duellman. Pp. 349-357, pl. 25, 2 figures in text. April 27, 1961.

9. Fish populations, following a drought, in the Neosho and Marais des Cygnes rivers of Kansas. By James Everett Deacon. Pp. 359-427, pls. 26-30, 3 figs. August 11, 1961.

10. Recent softshelled turtles of North America (family Trionychidae). By Robert G. Webb. Pp. 429-611, pls. 31-54, 24 figures in text. February 16, 1962.

Vol. 14. 1. Neotropical bats from western Mexico. By Sydney Anderson. Pp. 1-8. October 24, 1960.

2. Geographic variation in the harvest mouse, Reithrodontomys megalotis, on the central Great Plains and in adjacent regions. By J. Knox Jones, Jr., and B. Morsaloglu. Pp. 9-27, 1 figure in text. July 24, 1961.

3. Mammals of Mesa Verde National Park, Colorado. By Sydney Anderson. Pp. 29-67, pls. 1 and 2, 3 figures in text. July 24, 1961.

4. A new subspecies of the black myotis (bat) from eastern Mexico. By E. Raymond Hall and Ticul Alvarez. Pp. 69-72, 1 figure in text. December 29, 1961.

5. North American yellow bats, "Dasypterus," and a list of the named kinds of the genus Lasiurus Gray. By E. Raymond Hall and J. Knox Jones, Jr. Pp. 73-98, 4 figures in text. December 29, 1961.

6. Natural history of the brush mouse (Peromyscus boylii) in Kansas

with description of a new subspecies. By Charles A. Long. Pp. 99-111, 1 figure in text. December 29, 1961.

7. Taxonomic status of some mice of the Peromyscus boylii group in eastern Mexico, with description of a new subspecies. By Ticul Alvarez. Pp. 113-120, 1 figure in text. December 29, 1961.

8. A new subspecies of ground squirrel (Spermophilus spilosoma) from Tamaulipas, Mexico. By Ticul Alvarez. Pp. 121-124. March 7, 1962.

9. Taxonomic status of the free-tailed bat, Tadarida yucatanica Miller. By J. Knox Jones, Jr., and Ticul Alvarez. Pp. 125-133, 1 figure in text. March 7, 1962.

More numbers will appear in volume 14.

Vol. 15. 1. The amphibians and reptiles of Michoacan, Mexico. By William E. Duellman. Pp. 1-148, pls. 1-6, 11 figures in text. December 20, 1961.

2. Some reptiles and amphibians from Korea. By Robert G. Webb, J. Knox Jones, Jr., and George W. Byers. Pp. 149-173. January 31, 1962.

3. A new species of frog (Genus Tomodactylus) from western Mexico. By Robert G. Webb. Pp. 175-181, 1 figure in text. March 7, 1962.

More numbers will appear in volume 15.

THE END